MUSEUM of MONOCULTURE

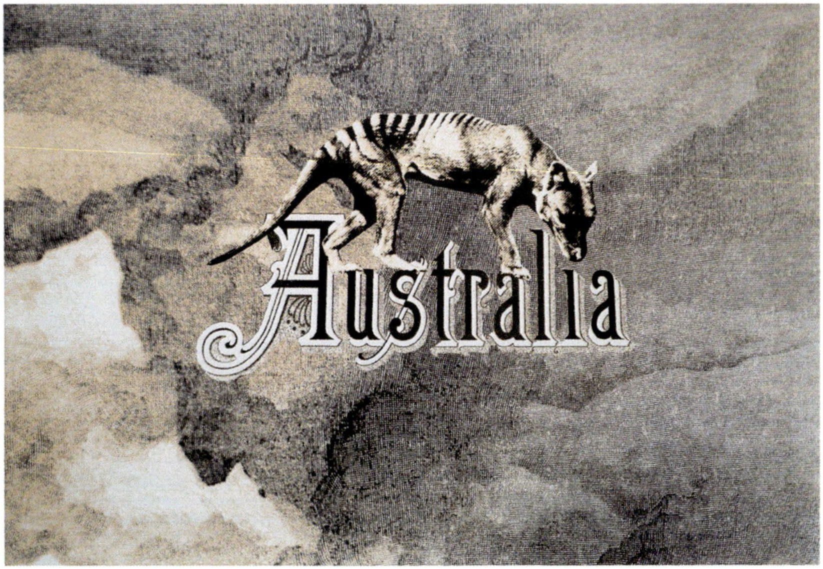

24 screenprints

Dudley House
ARTISTS ON VIEW

Justine Philip PhD

A c k n o w l e d g e m e n t s

Many thanks to Bendigo Venues and Events for hosting the exhibition with their Artists on View program. Thanks to Roland Breckwoldt and Professor Frank Uekötter for their essay contributions to this catalogue, and their valuable feedback. Huge thanks goes to Garth Maxwell who commissioned the first 4 screens, and to Mark Wilkinson and the Birmingham Printmakers for teaching me the ropes of the screen printing process. Thanks to Philip Doggett-Williams and the printmakers at Sunshine Print Artspace in Melbourne, an inspiring place just to be in. Thanks to my colleagues on the MaMoGH project 2022-2024; Kapil Subramanian, Volodymyr Kulikov and Gabriella Petrick, and to the University of Birmingham, Prof William Purkis and the School of History and Cultures for supporting my Honorary Research Fellowship. All thanks to Joni, Moira, Enua and Philip for their feedback, enthusiasm, editing skills, and patience. And to Prince, for accompanying me from France to the UK, and on to Melbourne and Bendigo. This is for you and all your messy, furry, feathered, noisy friends.

Now, let's go shoot the breeze.

Published October 2025 by the Museum of Monoculture,
Bendigo, Australia. Printed by Eureka Press Melbourne

Limited edition of 200

ISBN: 978-1-7643046-0-3

TENZING STUDIO

MUSEUM of MONOCULTURE

SUNSHINE PRINT ARTSPACE

Birmingham PRINTMAKERS

erc European Research Council

UNIVERSITY OF BIRMINGHAM

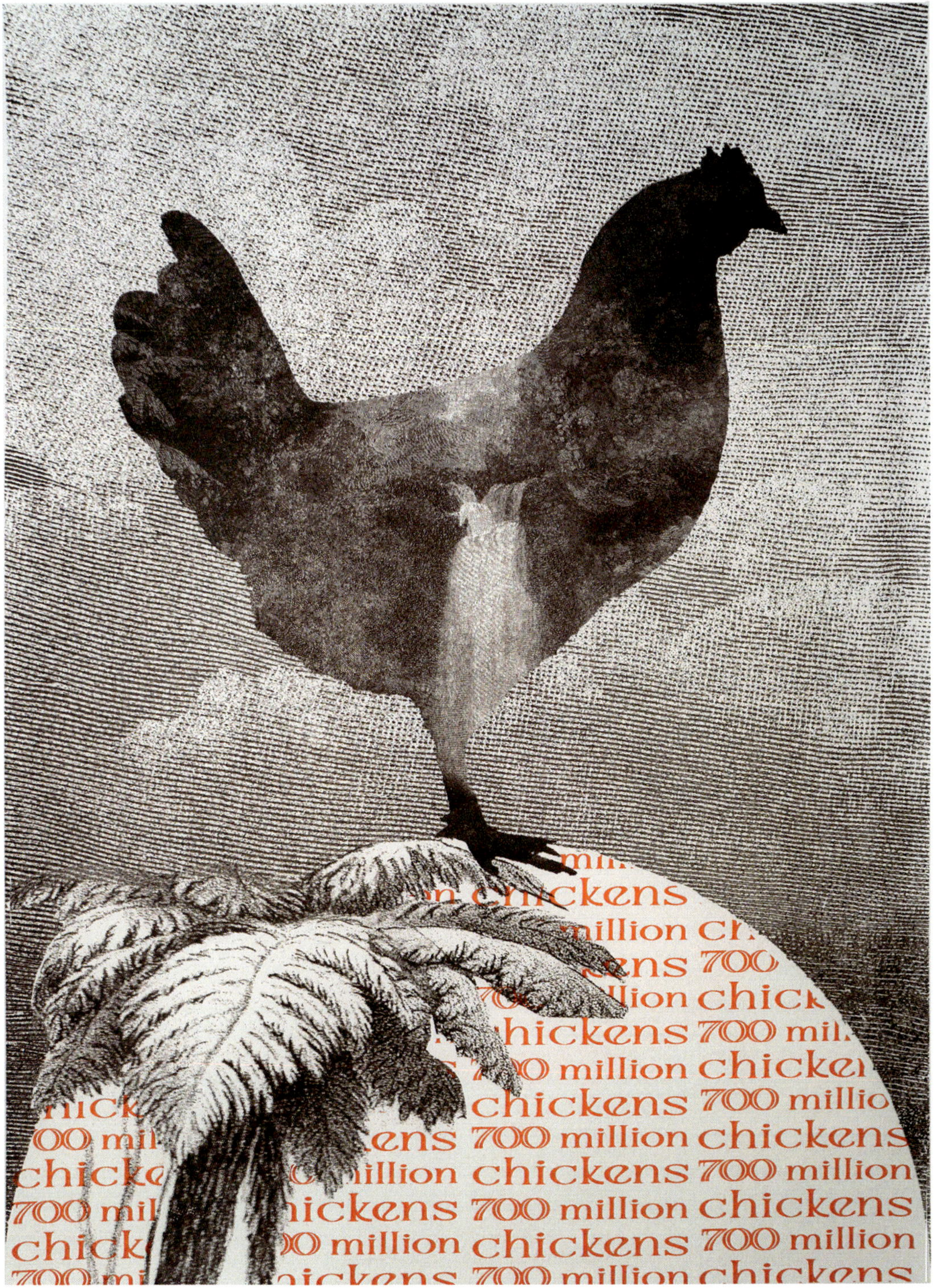

CONTENTS

Chicken

Photomontage
screenprint

Acrylic on
Fabriano
Rosaspina
50 x 70cm

Introduction

Monoculture (noun): ***The cultivation or exploitation of a single crop, or the maintenance of a single kind of animal, to the exclusion of others. [1].***

This exhibition explores how monoculture, as a defining cultural system, is shaping our economy, landscapes, and ecology. It features 24 original screenprints by Bendigo-based agricultural historian and printmaker, Justine Philip.

The series was inspired during Philip's work on the European Research Council (ERC) project *The Making of Monoculture. A Global History* (MaMoGH). Industrial agriculture once enjoyed a productive partnership with the visual arts well into the mid-20th century. Museums of Industry and Technology flourished, and artists were encouraged to record technological advances alongside images that celebrated the rural landscape and agricultural labour. Since the late 1940s, however, the hyper-productivity of modern food and fibre production has largely disappeared from public view.

Frank Uekötter's essay *A View to a Kill* examines the drivers behind this shift, while Roland Breckwoldt explores ancient traditions of representing animals as a way of revealing humanity's changing relationship with the natural world.

Monoculture's disregard for the biophysical limits of land, water, labour, and economics has profoundly shaped the lives of millions of animals in Australia and New Zealand. Rare and endangered wildlife, free-living animals labelled as pest species, and livestock are all entrapped within the system. The alarming pivot towards employing agrochemical technologies as frontline conservation tools—used to kill unwanted wildlife under the guise of protecting the rarest species—is arguably unconscionable.

Philip's works confront this reality through photomontage screenprints on paper (500mm x 700mm). Drawing inspiration from the protest poster and the 1920s designs of the British Empire Marketing Board, these images reveal how agricultural landscapes have been reframed, and how animal life has been rendered increasingly fragile within them.

[1] Oxford University Press. Oxford English Dictionary. Oxford University Press; 2020.

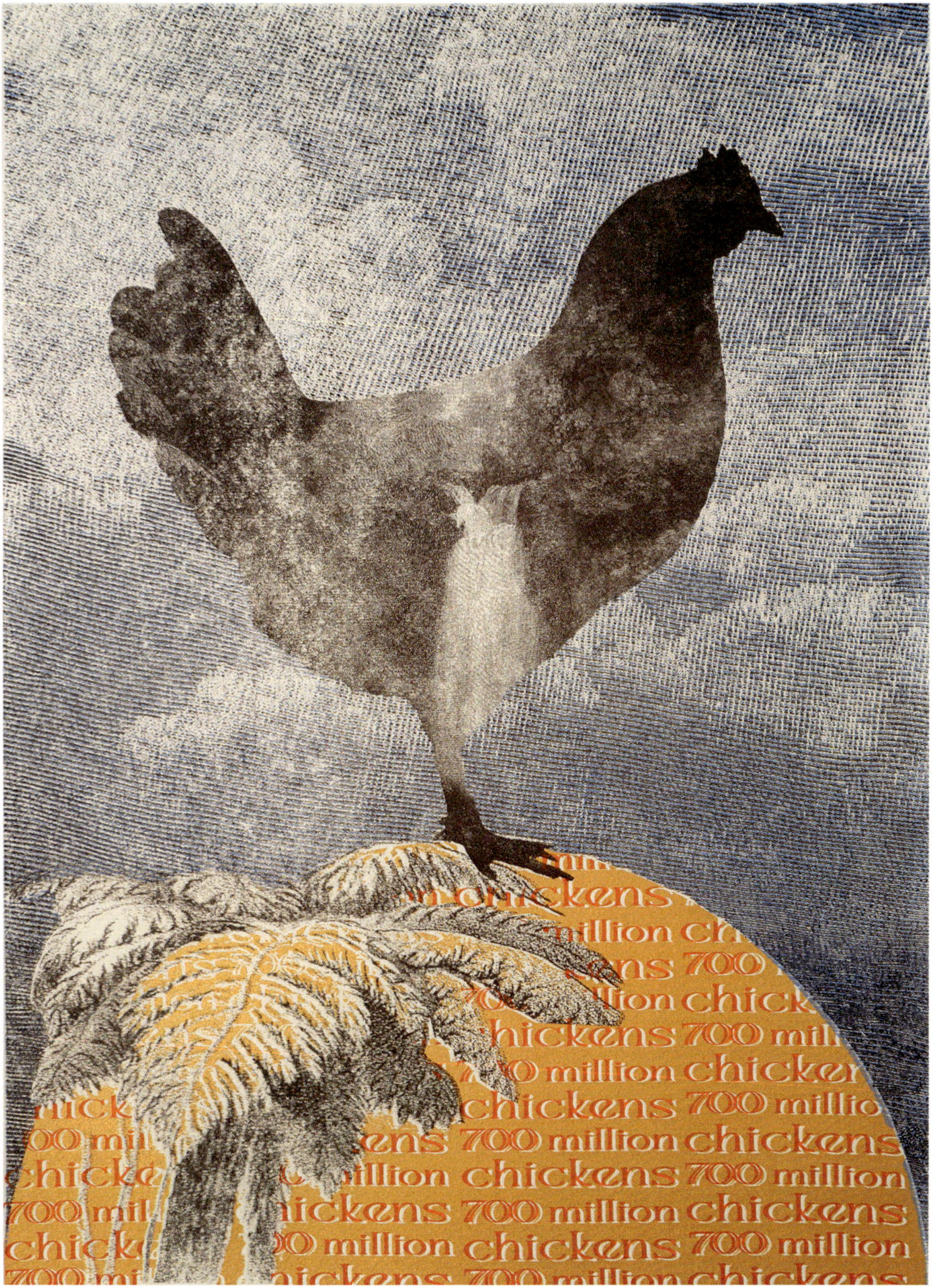

Shooting the Breeze

Photomontage screenprint

Acrylic on Fabriano Rosaspina
50 x 70cm

A View to a Kill

Professor Frank Uekötter

Monoculture is not a visual story, and it is not meant to be. Monocultures are supposed to deliver food, and do so in quantities that feed a global population of more than eight billion. Every agriculturalist can tell the story of the modern cornucopia: advances in science and technology allow a dwindling group of farmers to produce more food than ever. They are less eager to talk about the toll for people and places, and they rarely come acknowledge the gaping intellectual hole at the heart of monoculture. When it comes to modern farming, we are stuck with a project that we struggle to explain, let alone justify, in clear conceptual terms.

The drift towards monoculture is a truly global trend. Wherever farmers have entered the orbit of supercharged global capitalism, they have gravitated towards a reliance on one single commodity: in the industrialized North and the Global South, in fields, groves and barns, in bulk production and precious spices. Given this pervasive trend, one would expect that there was – and is – a robust case for monoculture, but there is none. We have plenty of conceptual and empirical evidence for biological diversity and nothing of that kind for the opposite: monoculture may well be the greatest project of global modernity that we have embarked upon without a decent paradigm. There is a pervasive speechlessness at the heart of monoculture, which may account for the remarkable lack of artistic visualization. How can we have pictures when we do not even have words?

Order in the Land

We have large fields with orderly rows of trees and plants, spaced for the large machines that rule in today's agriculture. Planting, spraying, harvesting –commodity chains are mechanized to an extent nowadays that many farm products remain untouched by human hands until they are consumed. There are precise numbers for the space that machines require, just as there are numbers for the perfect speed, the dose of chemicals, the depth of tillage and so forth. Quantification is the preferred language of the world of monoculture, and it is about more than technical needs. Numbers carry an aura of objectivity and precision, they invite comparisons and call for optimization, and they suggest to the mathematically inclined that there was order in the land.

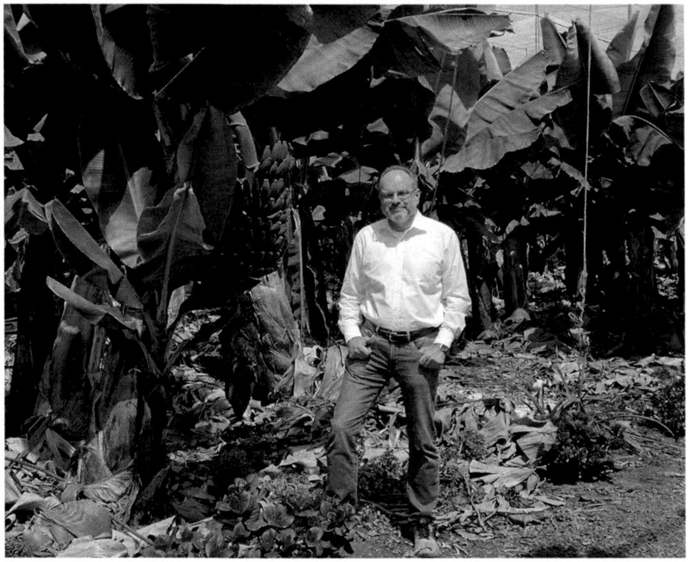
Frank Uekötter visiting a banana plantation on Teneriffa.
PHOTO: Frank Uekötter 2025.

A semblance of order was much desired in the modern food business. In the world of monoculture, the daily bread was struggling with all sorts of crises. A lot could go wrong on long commodity chains that spanned the planet. Consumers changed their tastes and preferences, and expanding supplies often led to overproduction and the collapse of prices. Workers could revolt or run away. Machines could break down for lack of energy or repairs. And then there was the specter of biological threats. Large stands of identical plants are the perfect breeding ground for pests and pathogens, and every monoculture faced a multitude of challenges from Mother Nature, each of which could literally eat up the biological essence of a farming system. Producers spent a lot of money on research and technology to keep these problems at bay, but even the most powerful companies faced limits when biological challenges endured. The infamous United Fruit Company spent some US$ 140 million between 1946 and 1956 to contain a devastating fungus that threatened their traditional Gros Michel banana variety, but it failed to deliver a satisfactory response. United Fruit only survived because it switched to the disease-resistant Cavendish banana, and because it had a powerful marketing apparatus that convinced consumers that the new variety, different in taste and outlook, was suddenly a real banana.[1]
The neat rows in fields and groves facilitate spraying against the biological troubles of the day, and sometimes they are also prevention strategies in their own right. Californian citrus farmers trim their trees into cubicle-sized units because it increases ventilation, which curbs the growth of fungi in brushwork.

Large stands of identical plants are the perfect breeding ground for pests and pathogens

And then there are the ethnic and political cleavages that call for perennial watchfulness. Many monocultures hinged on cheap non-white labor, but their recruitment triggered white anxieties, particularly when political power was not in the hands of white men. Gregg Mitman has argued for Liberia that the strict geometric patterns in Firestone's rubber plantations were a tool to placate the nerves of embattled ex-pats: "Order helped to rein in fears of chaos, confusion, and labor unrest that lived just beneath the surface of the idyllic world that Firestone sought to create and project for its white foreign staff."[2]

Battlefields

Spraying was not the only option when biology refused to go along with the wishes of the men of monoculture. In 1932, a response of choice was the seventh heavy battery of the Royal Australian Artillery, whose machine guns were supposed to eliminate the emu herds that damaged wheat fields in Western Australia.[3] The "Emu War" had a clear winner – few birds were hit, and quite a few emus survived catching a bullet –, but there was more to the campaign than a military folly. The fields and groves of monoculture were battlefields of sorts, and the struggle to keep production alive was remarkably similar in tools and mindsets to modern warfare. In both cases, the key challenge was operating complex technological systems in a hostile environment.

The difference was that we have plenty of pictures for modern warfare but few for the battle against biological threats. A crop duster plays a side role in Alfred Hitchcock's North by Northwest, but it targets Cary Grant rather than pests or pathogens, and it was hard to give a heroic spin to the daily battle out in the land. Killing pesky insects was rarely seen as more than an agronomic necessity, and few people had an idea of the complexity of the endeavor. Using the right chemical at the right time in the right dose was a task for professionals, but there were no popular narratives for their job. And who had a proper understanding of how deep plowing, a common way of curbing weed problems in the age of high-powered tractors, put soil structures and humus layers at risk?

The troubles of monoculture encouraged a take-no-prisoner's attitude. Producers sought responses to the many crises that surrounded them, and in making choices, they rarely thought about more than costs and benefits. Side effects were for others to worry about. When Queensland farmers found that their sugarcane fields turned acidic, they sought lime from the cheapest place that could deliver, which happened to be the Great Barrier Reef. Coral mining began in at least twelve areas around 1900 and continued over four decades.[4] The battlefields of monoculture extended far beyond the places of production.

Beyond Words

The Indian novelist Amitav Ghosh has drawn attention to the notable cultural inertia of the history of petroleum: it "produced scarcely a single work of note." In his judgment, the toll of our infatuation with fossil energy was beyond words: "The history of oil is a matter of embarrassment verging on the unspeakable, the pornographic."[5] One might say the same about modern agriculture if it were not for all the happy pictures that we have in our heads: bucolic landscapes, smiling farmers, delicious-looking food. Many of these pictures were the product of carefully crafted propaganda from agrobusiness. Others are mere leftovers from the bygone age of premodern farming, when most people lived in the countryside rather than cities. We have plenty of stories when it comes to our food, but they have as much to do with the reality of farming as James Bond with the daily routine of the British civil service.

Many studies have shown the social, economic, political and ecological costs of narrow agricultural specialization, and yet monoculture leaves a void in our cultural imagination. It is about thoughtlessness, about convenience, about the cultural hegemony of a carefree mass consumption paradigm that refuses to die. But maybe there is something else: a stubborn inability of our culture to connect the dots. Maybe it would be different if we had something visual that allowed us to recognize our long and fateful love affair with monoculture, and our own roles in that ongoing story. The field is wide open for artists with an independent mind, a concern for real-world issues, and the creativity to envision the invisible. And what would better show the power of the arts than guiding us towards a path that finally brings us to see the full picture.

References

[1] Stacy May, Galo Plaza, The United Fruit Company in Latin America (Washington DC: National Planning Association, 1958), p. 155; John Soluri, Banana Cultures: Agriculture, Consumption, and Environmental Change in Honduras and the United States (Austin: University of Texas Press, 2005), pp. 180-83.

[2] Gregg Mitman, Empire of Rubber: Firestone's Scramble for Land and Power in Liberia (New York: The New Press, 2021), p. 182.

[3] Murray Johnson, "'Feathered Foes': Soldier Settlers and Western Australia's 'Emu War' of 1932," Journal of Australian Studies 30 (2006), pp. 147-157.

[4] Cf. Ben Daley, Peter Griggs, "Mining the Reefs and Cays: Coral, Guano and Rock Phosphate Extraction in the Great Barrier Reef, Australia, 1844-1940," Environment and History 12 (2006), pp. 395-433.

[5] Amitav Ghosh, "Petrofiction: The Oil Encounter and the Novel," Imre Szemann, Dominic Boyer (eds.), Energy Humanities: An Anthology (Baltimore: Johns Hopkins University Press, 2017), pp. 431-440; p. 431.

Art and Agriculture

Roland Breckwold
Australian Author/Stockman

My first encounter with Justine was when I read her eloquent thesis on the dingo. I was set back on my heels when seeing how art can be a part of science. The intense effort and thought involved in her work was evident. No international collection was missed, no detail overlooked and compelling language within them was maintained.

So it is no surprise at all to learn that Justine applies her unique vision and linguistic skills to produce posters that remind us of the world we live in and its wonderful possibilities. This interest in finding visual expression to interrogate our relationship with our food resources has ancient roots. It is said that the first known artists carved designs in Blombos Cave in Africa about 73,000 years ago. Such is the urge within us to capture the subject of this very exhibition, 45,000 years ago the people who used Leang Tedongnge Cave on the Indonesian island of Sulawesi painted three pigs on its limestone walls.

Perhaps at a similar time, the Neanderthals entered deep caves to create their own records, forming a ceremonial circle of broken stalagmites and using shells as ornaments or containers for pigments. Neandertals used the feathers of birds such as the Eurasian black vulture, golden eagle, red-footed falcon and common wood pigeon, changing the structure of their feet to adorn themselves or construct symbolic arrangements as long ago as 44,000 years ago.

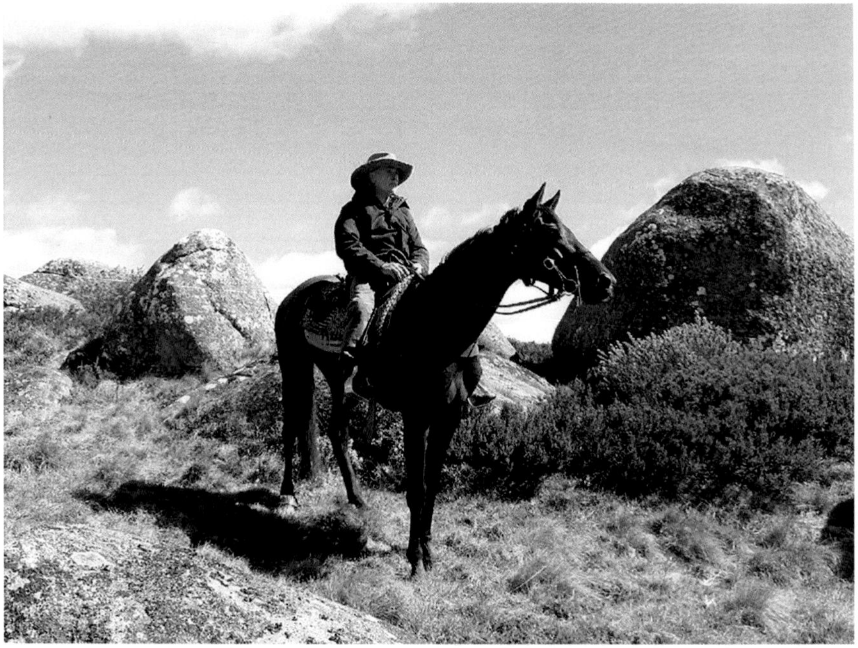

Roland Breckwoldt PHOTO: Sarah Carlill 2024

The delicately beautiful art that comes alive before your eyes in deep galleries such as France's Lascaux Cave are the creative drive of unknown artists. They are dated to about 20,000 years ago invariably described as the work of 'early palaeolithic people' rather than choosing between Neandertal or Sapiens or any combination of the two. The hundreds of paintings within the Lascaux Cave are the first example of art showing the internal organs and physiology of the wild animals early artists preyed upon. One such work depicts a bison spilling its innards after being speared, with a bird-headed human lying between it and a decomposing rhinoceros. Analysts remind us that while cave paintings delight the eye and the imagination, interpreting its reason and intent is another matter.

In the Australian context, Indigenous Australians have recorded their hunting practices in cave paintings across the Australian continent for millennia.

So, dear observer, as you view this work by Justine, please consider it thoughtfully. This is different, Justine possesses an expansive repertoire of words and images that launch you into the beauty of the world around us and its incredible depth.

Monoculture and Art

Justine Philip

This series of 24 screen prints emerged out of my research on the MaMoGH project, **The Making of Monoculture. A Global History** at the University of Birmingham, UK 2022-2024. It was a visceral response to the study of modern agriculture; to the history, technology and *thinking* behind the mechanics of our giant food and fibre production systems.

I wrote an article for the independent news source *The Conversation*, raising concerns about the pivot towards redeploying some of the most aggressive agrochemical technologies as frontline conservation tools in New Zealand and Australia. Here, they are aimed at removing a select group of introduced species—wild animals thriving in and around the highly disturbed landscapes designated for livestock or for conservation. The article was approved by my university, all the way up to the editor's desk in New Zealand, where it was ultimately refused publication. In response, I printed my first protest poster, *Te Wāhipounamu*, as a way of giving voice to these issues—to question the ethics behind our food choices, and the poor treatment of the treasured, the undesirable, and the forgotten species caught up in the wheels of industrial agriculture.

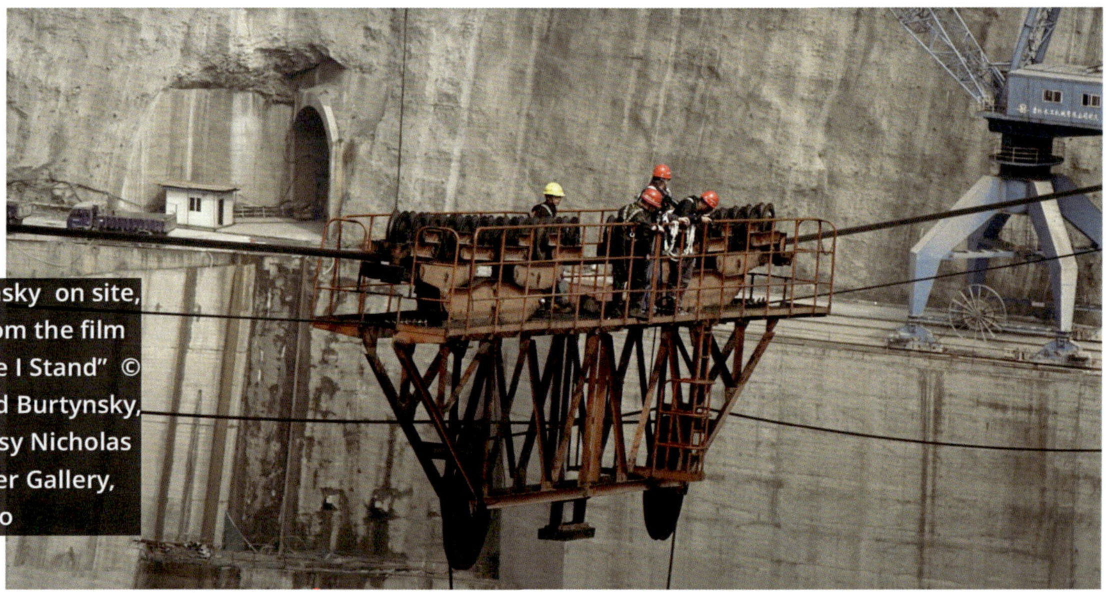

Burtynsky on site, still from the film "Where I Stand" © Edward Burtynsky, courtesy Nicholas Metivier Gallery, Toronto

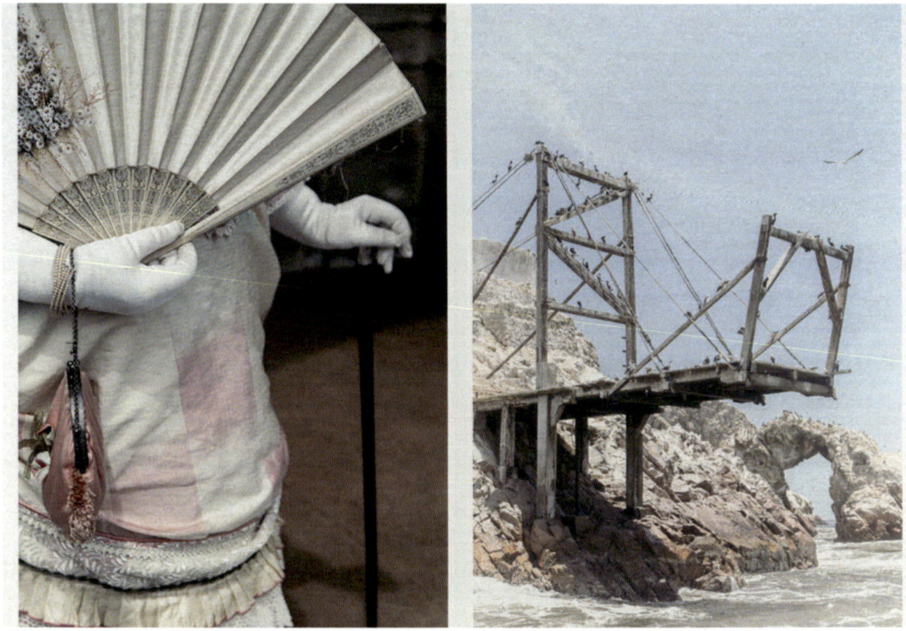

Olli Heliman"s book *$tow High in Transit ($hit)* provides a visual history of the Peruvian guano trade that fertilised the fields of the British colonies for three decades of the 19th century. © Olli Helimann

The lack of visual archives documenting these monocultural operations is problematic. Monoculture has become our primary method of food production, yet it rarely features as a subject in art galleries or museum displays. Exhibitions celebrating rural and agrarian life were popular before the mid-20th century, but their traditional status declined after World War II. Since then, the field has become increasingly volatile. Where artists were once encouraged—and often commissioned by industry—to document the latest technological advances in the production of food and fibre, many modern processes now operate under a veil of secrecy. This is particularly true of industrial livestock facilities and their vampiric processes. Australia raises and consumes 700 million chickens each year. Figures like that give pause for reflection—where are they all?

One artist who has risen to the challenge of recording our modern industrial sites at scale, is Canadian photographer Edward Burtynsky. He has spent the past four decades building an extraordinary archive of humanities engineering excesses. The images capture the true scale of industrial extraction in all its grandeur and terror, alongside the impact of the monstrously invasive agricultural systems that have transformed water, landscapes and human lives. Another is photographic artist Professor Olli Heliman, his photos of industry and wealth are juxtaposed side by side, exposing the beautiful excesses of colonisation next to the wreckage of the landscapes and lives left in its wake.

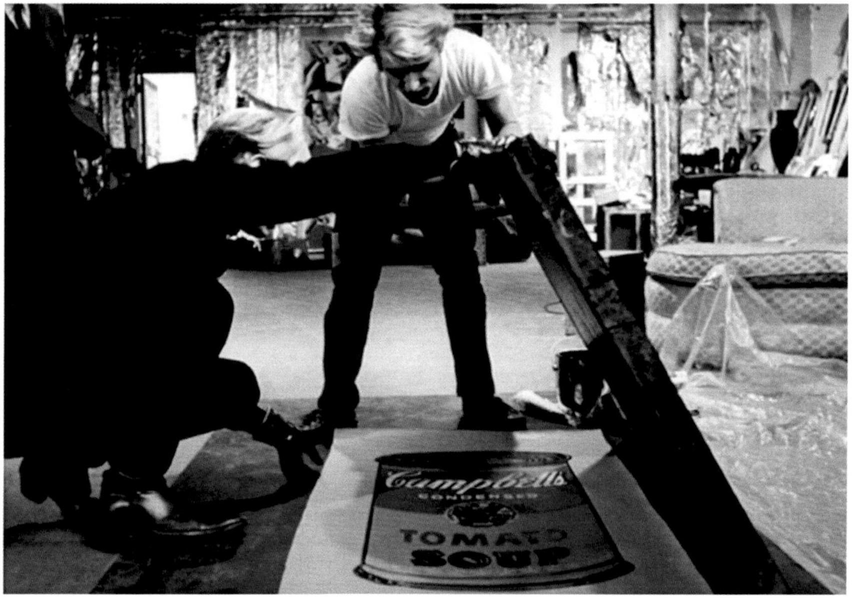

Andy Warhol printing silkscreens in his New York Factory 1960s.

These artists invite us into a conversation with technology, presenting the immense scale and productivity of modern engineering projects, and the audacity of the architects of its design. They recognise the costs to the natural world in the process. The role of the artist in this space can be to simply take a step back, and to just present the images, without flinching. Screen printing is a suitably industrial system for the task, providing a method of disseminating information with duplication at its core. I learnt the ropes at Birmingham Printmakers in the UK, where the shared studio space is popular with street artists and artisans. I completed all the New Zealand posters there in 2024, working in their studio on the third floor of an old factory building, in an industrial corner of the city centre.

In the 1960s, pop artists took screen printing to the galleries as they grappled with the symbols of mass consumption and industrialisation that engulfed them. Andy Warhol's first solo exhibition – featuring screen prints of 32 Campbell's soup cans, opened at Ferus Gallery on 9 July 1962. At 9am that same morning, the United States conducted "Starfish Prime", a high-altitude nuclear test, that detonated over the Pacific Ocean lighting up the night sky from Australia to Hawaii. The first Walmart had opened in Arkansas, US just seven days earlier.

In 2003, Banksy's exhibition "Turf War" opened in East London (18 July) drawing on Warhol's work of painted pig and cow portraits, to examine our complex relationship to the animals caught up in industrial agriculture. The works featured live farm animals covered in painted symbols and slogans. I have borrowed the silhouette from Banksy's work *Fuck Pigs* to construct the screens depicting the Ezhou pig towers in China. The Exhou Pig Towers were designed by a Chinese concrete construction company looking for a new direction to expand their ailing business, and they engineered the world's largest CAFO – Concentrated Animal Feeding Operation, completed in 2022. At 26 stories high, they produce over 1.2 million pigs each year for slaughter. The toll of these operations at scale on the environment, has been enough to bring down entire economies – see *Communist Pigs: An Animal History of East Germany's Rise and Fall*, by Thomas Fleischman (2022).

Aerial baiting features prominently in my research and in my screens, a technology that was invented in Australia the 1940s to protect agricultural lands from pest species, targeting both herbivores and predators. It enables the aerial broadcast of poison baits across vast, inaccessible areas of country – and operates far from the public gaze. The scale of these operations has been astonishing. In many years since 1947, Australia news media has covered articles announcing the drop of "One Million Baits" over public and private lands to tackle pest species, with the war rhetoric leaving little room of reflection. One of Australia's primary targets is the dingo, the country's top order predator. They remain the target of eradication campaigns today in the drylands – sheep country, bypassing all cultural and ecological considerations.

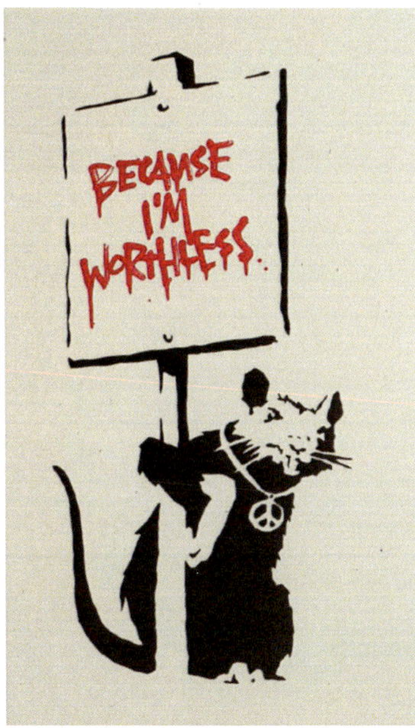

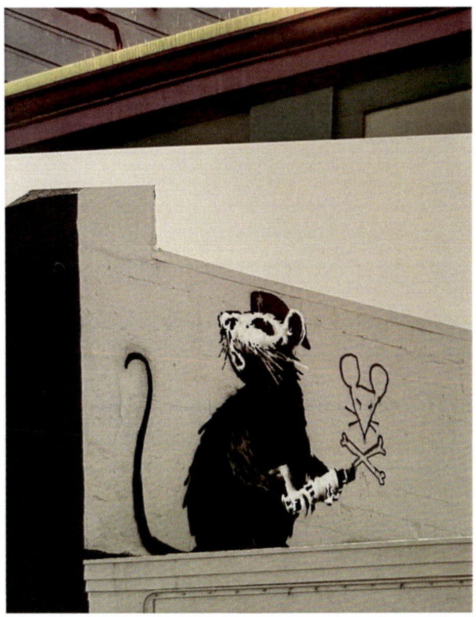

Bristol street artist Banksy on pest control.

Underground mutton (wild rabbit) provided Australia with a plentiful and reliable food resource from the 1890s until 1950, and supported an industry that flourished despite drought, war and economic downturns. The industry expanded in decades where the sheep industry contracted, and provided a good living for those without land title. There were jobs in the canning factories and freezing works, tanners and street stalls, hatters and furriers. By 1929 the rabbit industry was the largest employer of labour in Australia. Rabbits transformed the social structure of Australian society and threatened to destabilise it – Farm labour was hard to secure as there was more money to be made in rabbiting. rabbits were eating grass that Pastoralists considered belonged to them for their livestock.

An estimated four billion rabbit skins were exported between 1904 and 1947, and 27 million rabbits consumed by Australians each year during the 1940s. During World War 2, Australian soldiers marched into battle wearing slouch hats made of rabbit skins. Australia produced ≈5,5 million hats during the war at ten rabbits per hat,

zoologist Francis Ratcliff (1956) described the rabbit as a divider of classes in Australia: landholders (farmers and graziers) vs. the rural employees and "freelance" workers.

The escape of myxomatosis from research trials in 1949 not only decimated the rabbit industry and the livelihoods of thousands of Australians, it went on to destroy wild and domestic populations of rabbits across France, wider Europe and the United Kingdom. It was a brutal and cruel disease. Queen Elizabeth and Winston Churchill both spoke out against it. The RSPCA lead mercy squads to put the animals out of their misery _ avoiding the 2 weeks of intense suffering before death by myxomatosis. Britain was still on war rations, so the loss of their wild rabbits impacted millions.

In 1956, Francis Radcliffe, head of the (C)SIRO Wildlife Survey Section and responsible for the myxomatosis trials, was summoned to Edinburgh to attend the International Union for the Conservation of Nature conference on **The Biological Consequences due to the Presence ofMyxomatosis** (Edinburgh 1956). The 240 delegates in attendance had conducted detailed studies of the impact of myxomatosis on biodiversity across Europe and Britain. This included an immense loss of prey species, changes to vegetation communities, prey-switch and its consequences, the extinction of the Large Blue Butterfly (Maculinea arion) in the UK (reintroduced in 1983 from sweden). The impacts were in many cases, irreversible. Foxes increased attachs on livestock; weasles and stoats consumed many more eggs and small birds; cats turned to attacking poultry; breeding sucess of owls and other birds of prey plummeted. Rabbit fur had been used to construct nests, with small bird species having to construct inferior dwellings instead made of feathers, scraps of wool, vegetation. Some Plant species disappeared, gorse and brambles increased rapidly, forests changed in composistion.

Francis Ratcliffe's contribution to the record was printed in the conference publication **La terre et la vie** (1956)3 "Myxomatosis "escaped" from one of our test sites in the Murray Valley in 1950" said Ratcliffe. He summarised that "no special observations of experiments have been undertaken to determine the nature of any changes in species composition and balance". The reasoning was that it was "not profitable or practicable" to monitor native pastures in Australia, and there were so few native species left on agricultural lands that they warranted no special consideration.

In New Zealand Myxomatosis proved ineffective, so in 1958 the Government introduced a "Killer Policy" with The Rabbit Board undertaking destruction of all rabbits on properties across the country, funded by government subsidies and landholder rabbit tax. They used poisons including phosphorus, arsenic, sodium fluoracetate (1080), aerial baiting, dogging, night-shooting, fumigation, ripping of warrens, and explosives to bring the population under control while banning the rabbit industry.

The human population in Australia and New Zealand had to rapidly adjust,, and the chicken industry stepped up to fill the void. Industry sources estimate that three million meat chickens were produced in Australia in 1951. That number today is 700 million, each year, from vast, polluting monocultures that challenge the pillars of environmental ethics, animal ethics, and economic theory. They externalise key vulnerabilities, ignoring the biophysical limits of land, water, labour and capital. monoculture operates outside the known limits of economic theory and biological tolerance. The eradication of species designated as pests creats a waste product our of valuable resorces, and the methods use to eradicate them - the biological warfare, pesticides, lethat traps etc. destroy entire ecosystems.

RIGHT: Text behind RABBIT montage

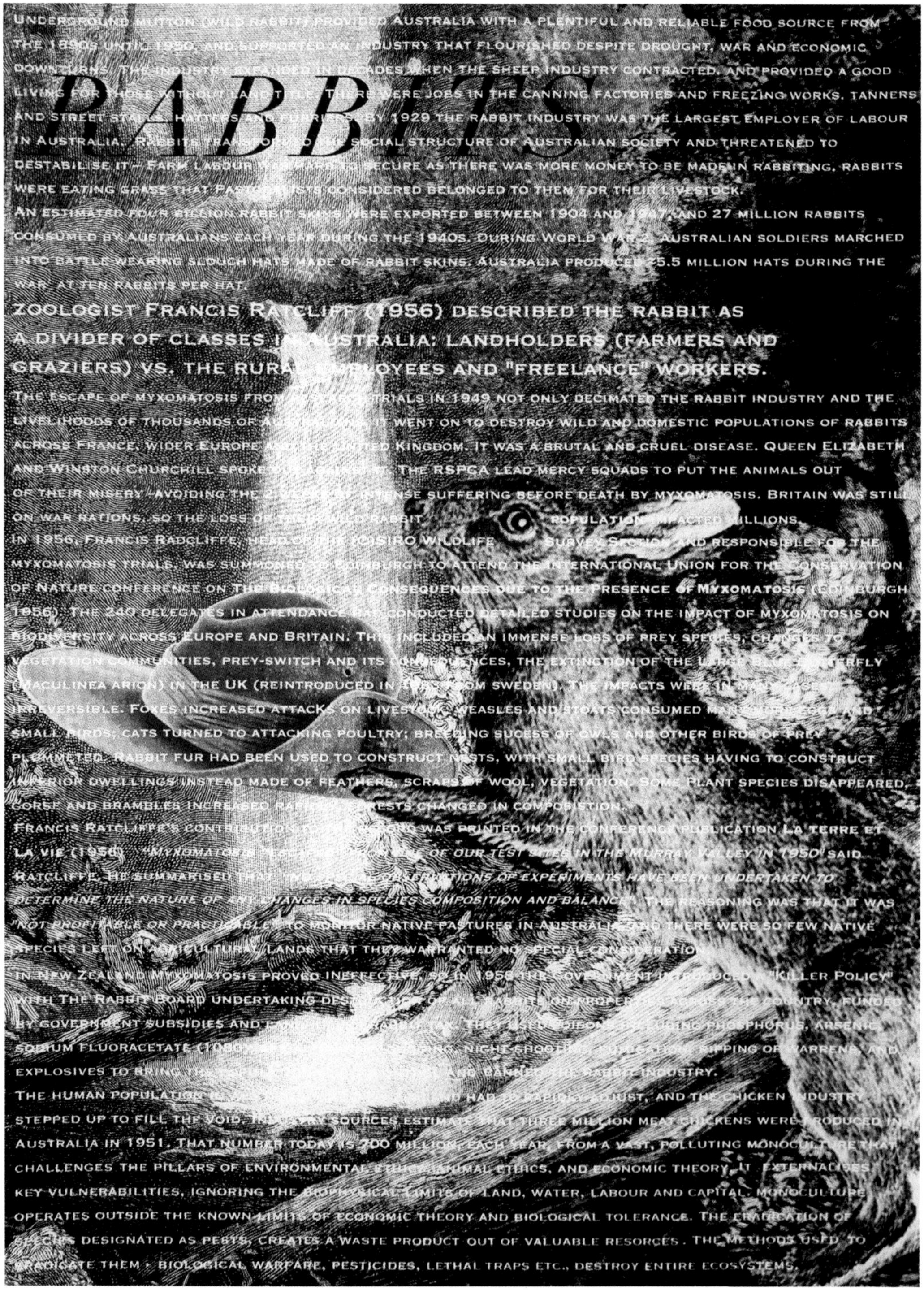

Rabbits 2025

Photomontage screenprint

Acrylic on Fabriano Rosaspina
50 x 70cm

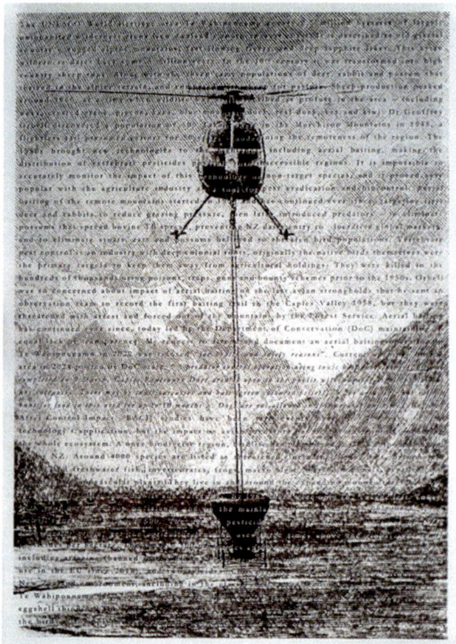

RIGHT:
Text behind image
Te Wāhipounamu

The UNESCO World Heritage area Te Wāhipounamu covers 2.5 million hectares of largely uninhabited wilderness along New Zealand's southern west coast. It is described as old glacial country – rugged alpine mountains, fast flowing rivers, and deep sapphire lakes. This avian wilderness dates back some 5 million years. In the 19th century it was transformed into high country sheep runs. Along with the sheep came populations of deer, rabbit and possum that thrived in the difficult terrain, and competed with sheep for pasture. Sheep production peaked around 1950, at a time when wildlife was described as profuse in the area – including weka, crested grebe, pigeon, kaka, blue mountain duck, teal duck, kea and kiwi. Dr Geoffrey Orbell discovered the lost Takahe here in The Murchison Mountains in 1948. a flightless rail presumed extinct for 50 years. Orbell uncovered a population of 500 surviving in the remote mountains. The 1950s brought new technologies to the highlands, including aerial baiting. It is impossible to accurately monitor the impact of this technology on non-target species, however it proved very popular with the agriculture industry interested in pest eradication and biocontrol. Aerial baiting of the remote mountains started in 1958 and has continued ever since, targeting first deer and rabbits to reduce grazing pressure, then later introduced predators – possum, stoat, rats, to protect the ailing bird populations. Vertebrate pest control is an industry with deep colonial roots, originally the native birds themselves were targets to keep them away from agricultural holdings. Hundreds of thousands were killed using poisons, traps, guns and bounty schemes. Orbell was so concerned about impact of aerial baiting on the avian community in Caples Valley that he sent an observation team in to record the first trial, but they were threatened with arrest and forced out of the mountains by the Forest Service. Aerial baiting has continued ever since, today led by the Department of Conservation upholding an equal lack of transparency. My request to attend and document an aerial baiting operation in Te Wāhipounamu in 2022 was refused "for health and safety reasons". Before-After-Control-Impact (BACI) studies have been rare events over the 70 years of the technology's application. A once biodiverse region, birdlife has plummeted in Te Wāhipounamu and across NZ. Around 4000 species are listed as threatened, including birds, bats, bryophytes, marine and freshwater fish, invertebrates, frogs, macro algae, marine mammals, reptiles, and vascular and nonvascular plants. They live in and around the increasingly contaminated soils and waterways. By 1982 there were only 182 Takahe left. Today, with sanctuaries set up on islands and on the mainland, the population is just back up to Orbell's original tally of 500. The use of pesticides is never cited as a driver of NZ's biodiversity crisis, yet the agricultural nation uses three times above the global average of pesticides per hectare. They have 3,500 pesticides approved for use by their industry run EPA, including atrazine (banned in the EU in 2003), neonicotinoids (completely banned in outdoor use in the EU since 2018), and two avicides. Chlorpyrifos is described as ubiquitous in the New Zealand environment, including in the glacial meltwaters feeding ethereal Lake Brewster at 1700m above sea level. Like DDT (not banned in NZ till 1989) it causes eggshell thinning. NZ's plan to save the birds is not to clean up its act, but to kill all its introduced predators – applying more vertebrate pesticides in the name of biosecurity and conservation.

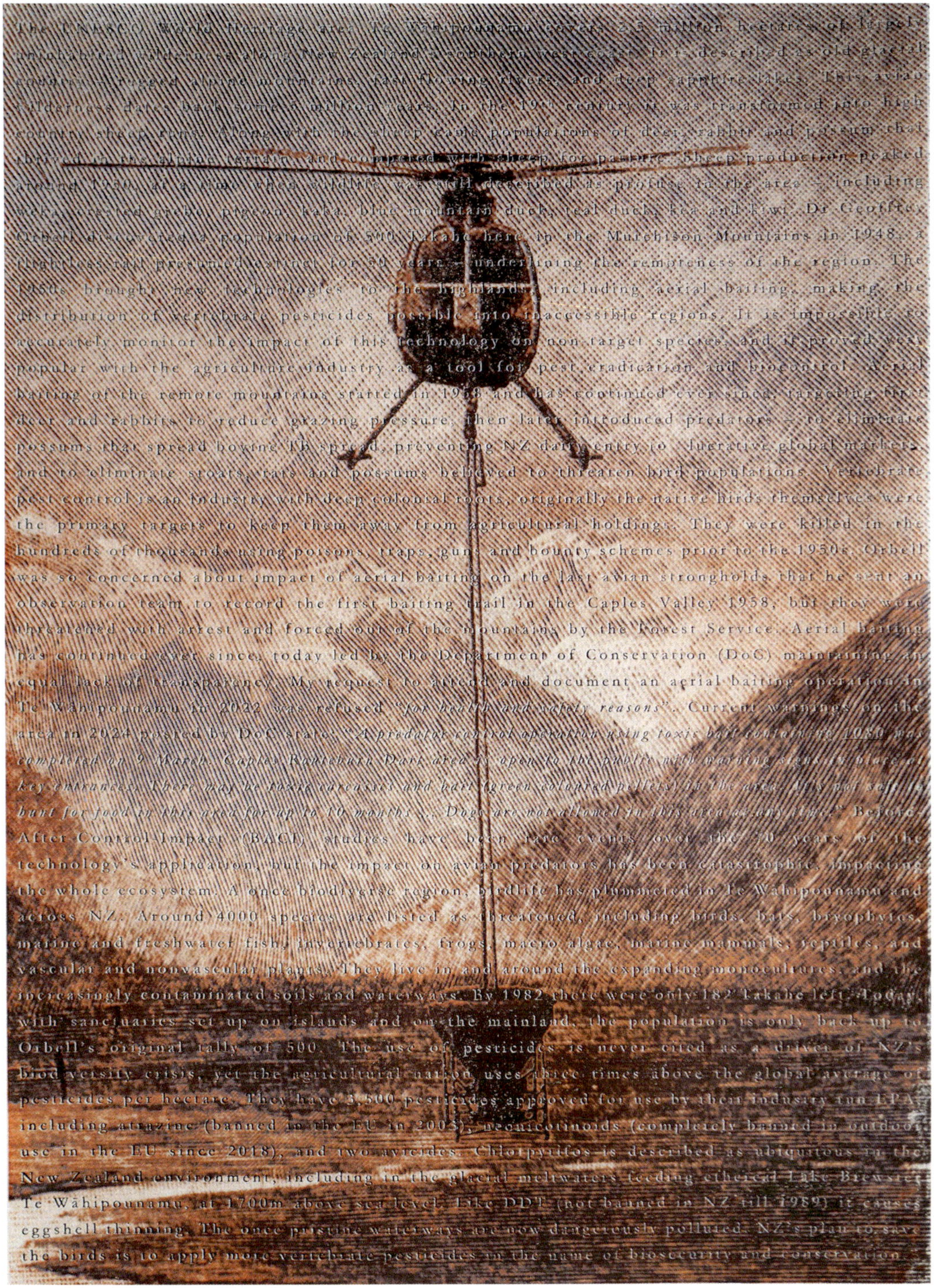

Te Wāhipounamu

Photomontage
screenprint

Acrylic on Fabriano
Rosaspina
50 x 70cm

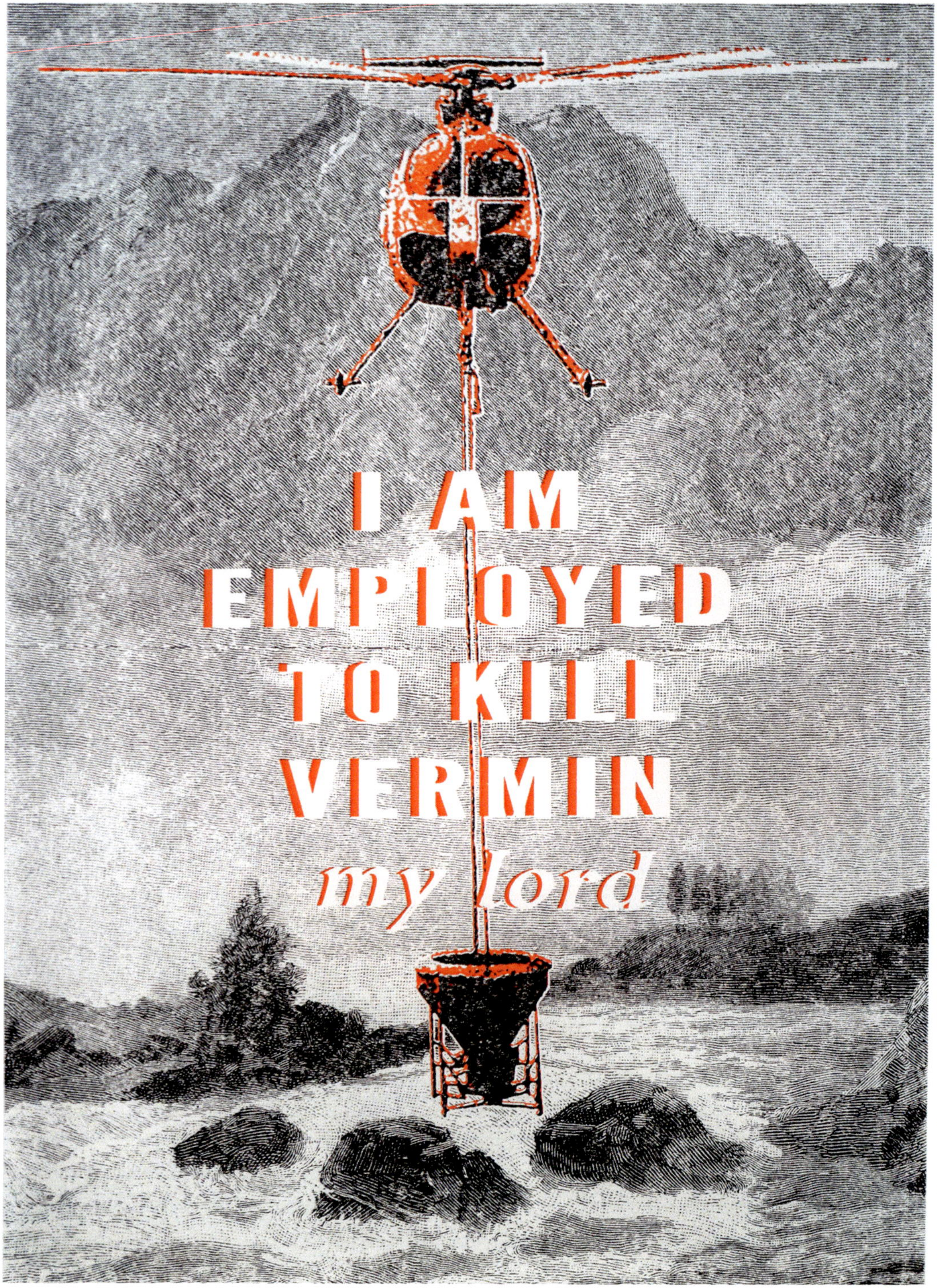

The Gamekeeper

Photomontage
screenprint

Acrylic on Fabriano
Rosaspina
50 x 70cm

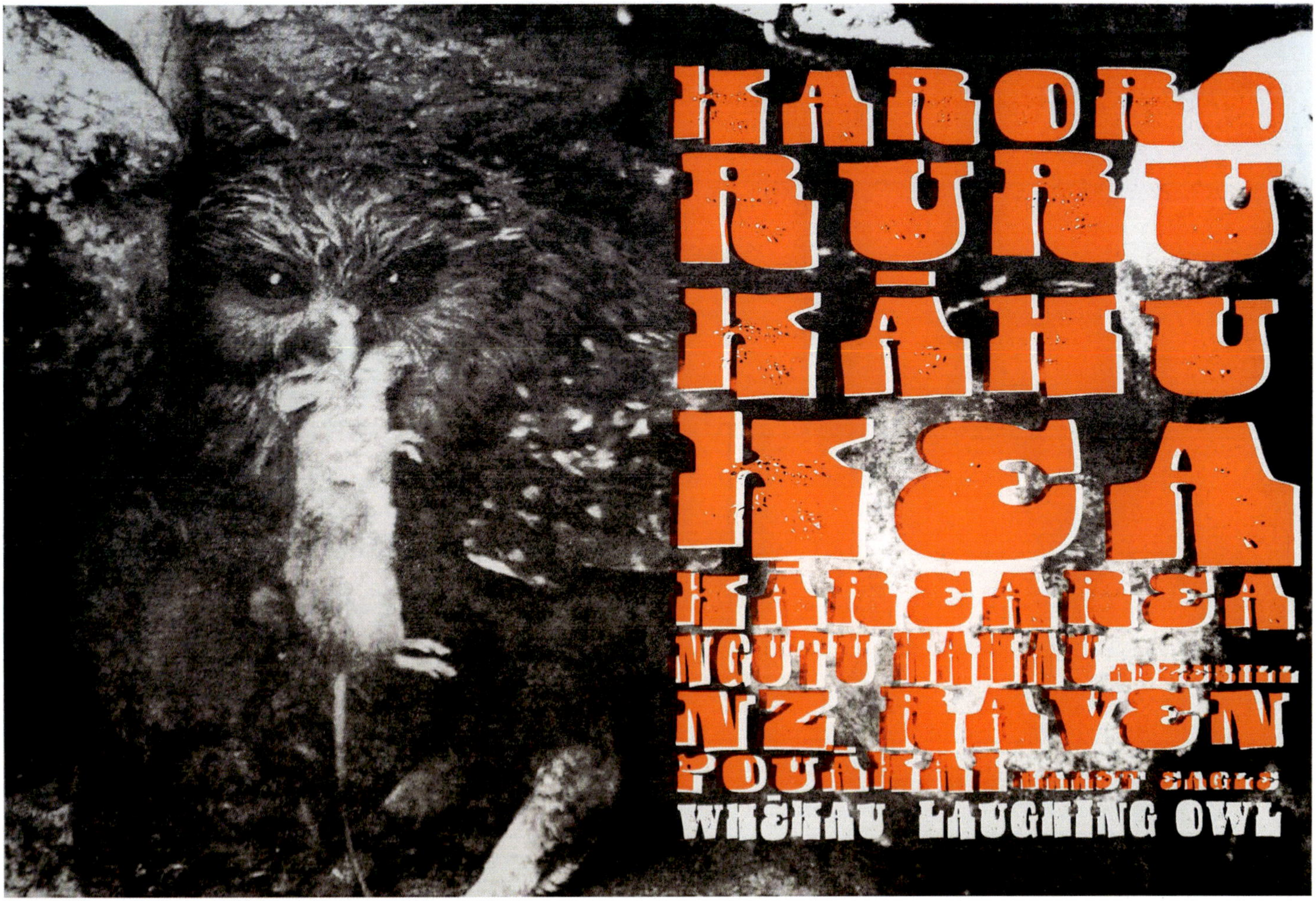

It is rare to find photographs of extinct species in the wild. This image was taken of the Whēkau in its nest - a cavity under a limestone boulder at Raincliff Station, South Canterbury 1909, by brothers Cuthbert and Oliver Parr.
The names printed include some of NZ's native predator birds; some extinct, others endangered and in the firing line with the agrochemical conservation project Predator Free NZ 2050.

Whēkau

Photomontage
screenprint

Acrylic on Fabriano
Rosaspina
50 x 70cm

South Island Kōkako

Photomontage screenprint

Acrylic on Fabriano Rosaspina
50 x 70cm

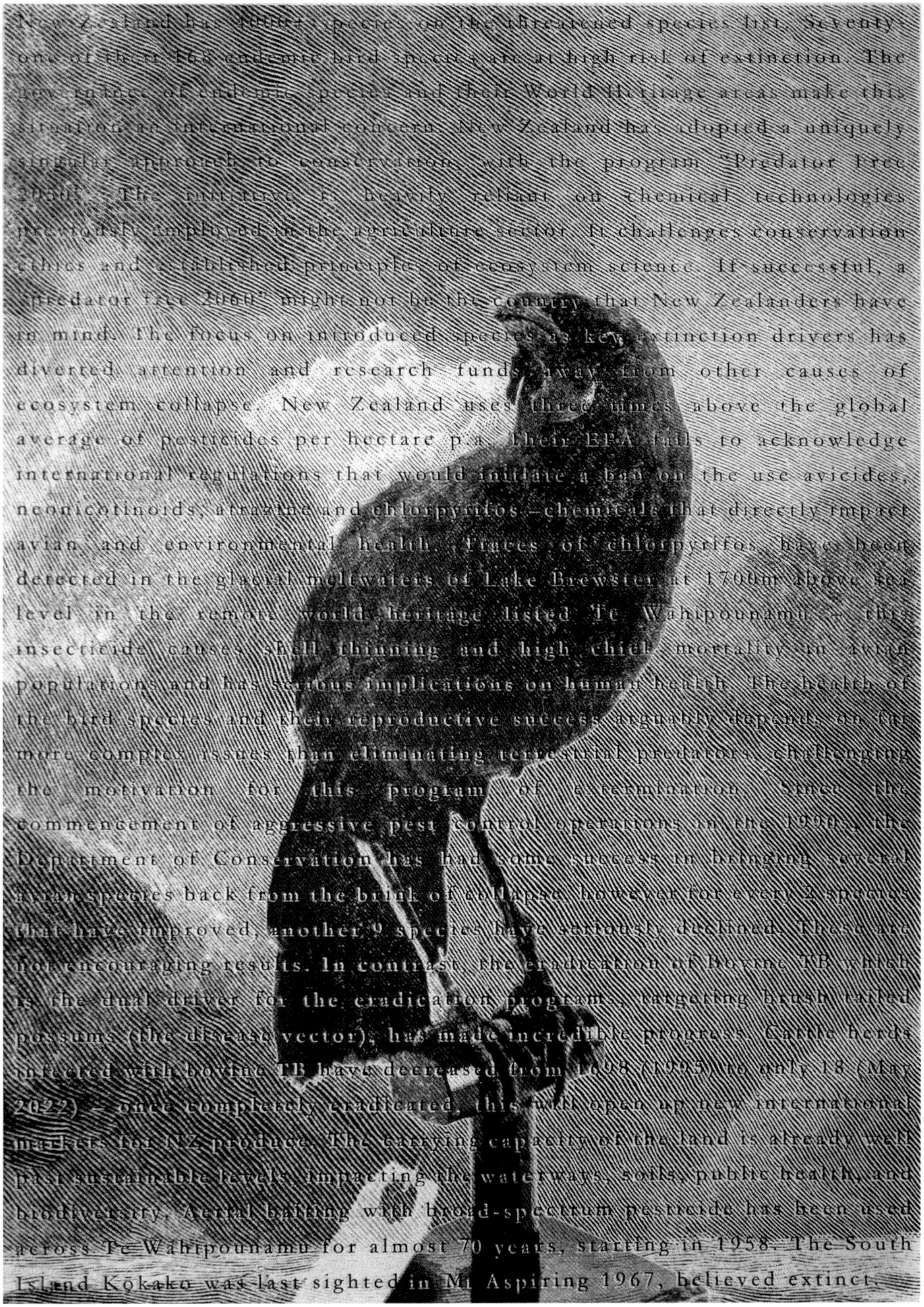
New Zealand has 4000+ species on the threatened species list. Seventy-one of their 168 endemic bird species are at high risk of extinction. The governance of endemic species and their World Heritage areas make this situation an international concern. New Zealand has adopted a uniquely singular approach to conservation, with the program "Predator Free 2050". The initiative is heavily reliant on chemical technologies previously employed in the agriculture sector. It challenges conservation ethics and established principles of ecosystem science. If successful, a "predator free 2060" might not be the country that New Zealanders have in mind. The focus on introduced species as key extinction drivers has diverted attention and research funds away from other causes of ecosystem collapse. New Zealand uses three times above the global average of pesticides per hectare p.a. Their EPA fails to acknowledge international regulations that would initiate a ban on the use avicides, neonicotinoids, atrazine and chlorpyrifos – chemicals that directly impact avian and environmental health. Traces of chlorpyrifos have been detected in the glacial meltwaters of Lake Brewster at 1700m above sea level in the remote world heritage listed Te Wāhipounamu – this insecticide causes shell thinning and high chick mortality in avian populations and has serious implications on human health. The health of the bird species and their reproductive success arguably depends on far more complex issues than eliminating terrestrial predators, challenging the motivation for this program of extermination. Since the commencement of aggressive pest control operations in the 1990s, the Department of Conservation has had some success in bringing several avian species back from the brink of collapse, however for every 2 species that have improved, another 9 species have seriously declined. These are not encouraging results. In contrast, the eradication of Bovine TB which is the dual driver for the eradication programs, targeting brush tailed possums (the disease vector), has made incredible progress. Cattle herds infected with bovine TB have decreased from 1698 (1995) to only 18 (May 2022) – once completely eradicated, this will open up new international markets for NZ produce. The carrying capacity of the land is already well past sustainable levels, impacting the waterways, soils, public health, and biodiversity. Aerial baiting with broad-spectrum pesticide has been used across Te Wāhipounamu for almost 70 years, starting in 1958. The South Island Kōkako was last sighted in Mt Aspiring 1967, believed extinct.

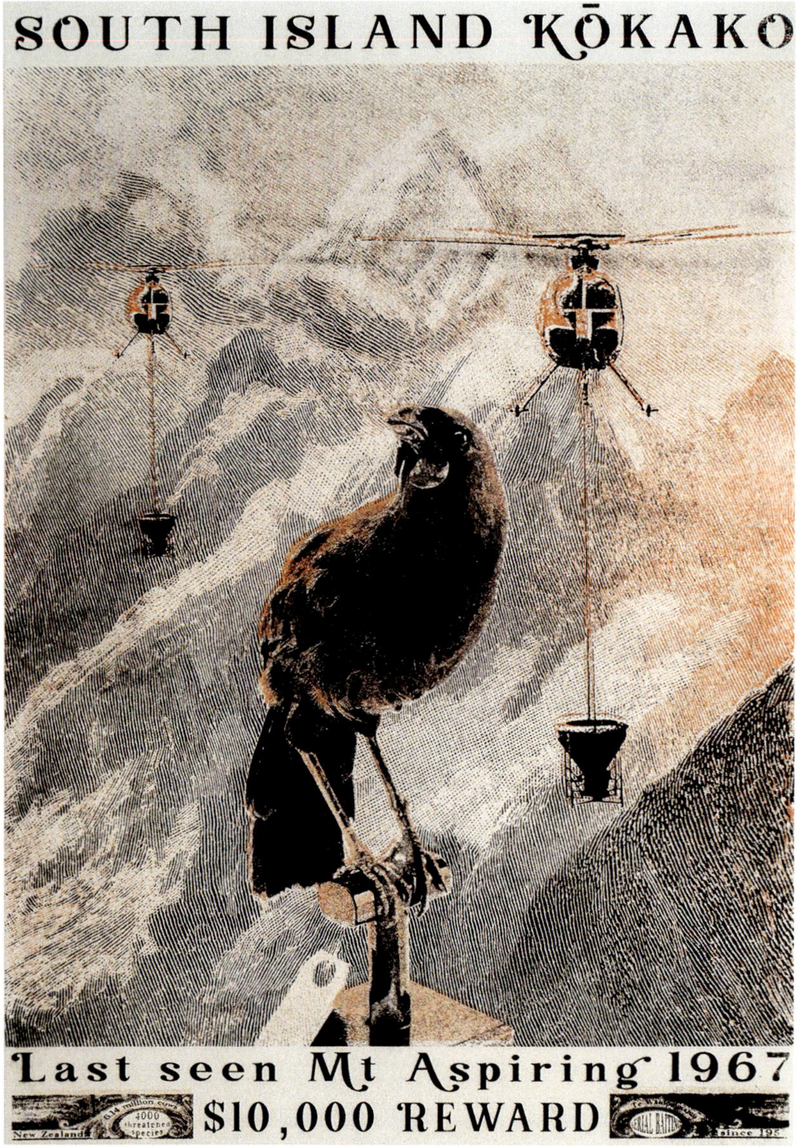

South Island Kōkako $10,000 Reward

Photomontage screenprint

Acrylic on Fabriano Rosaspina
50 x 70cm

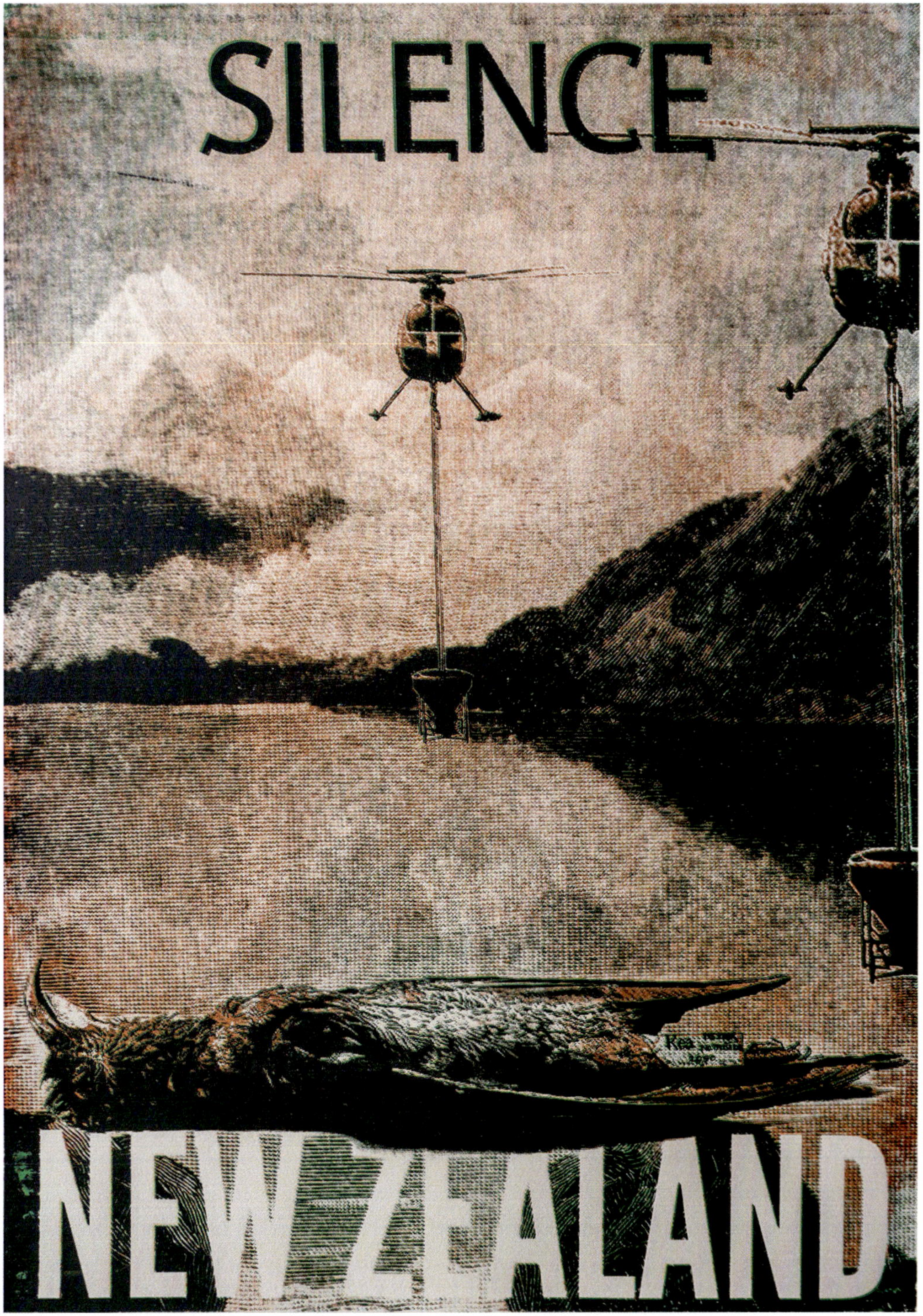

Kea

Photomontage
screenprint

Acrylic on Fabriano
Rosaspina
50 x 70cm

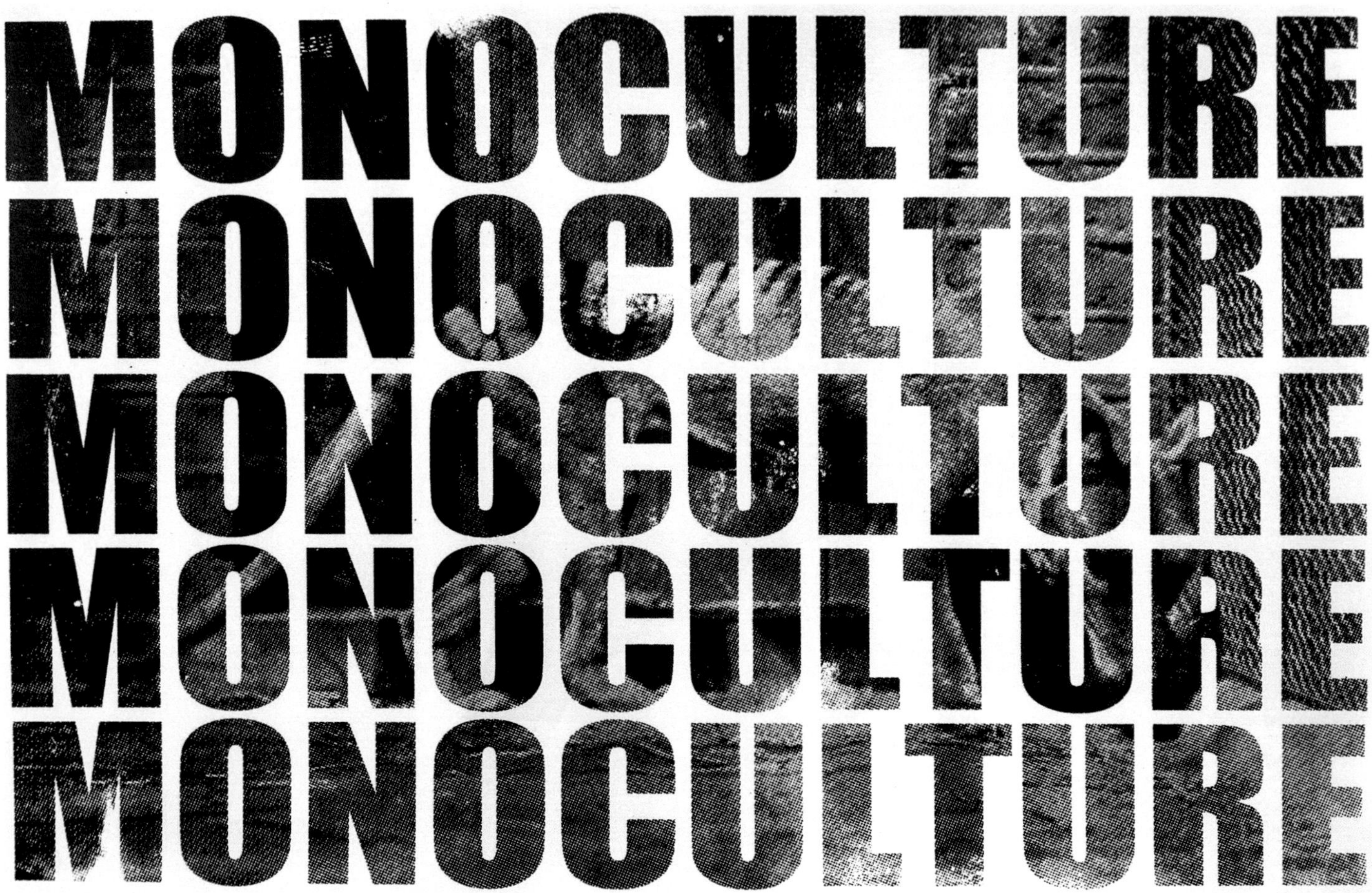

Monoculture

Photomontage
screenprint

Acrylic on
Snowdon

50 x 70cm

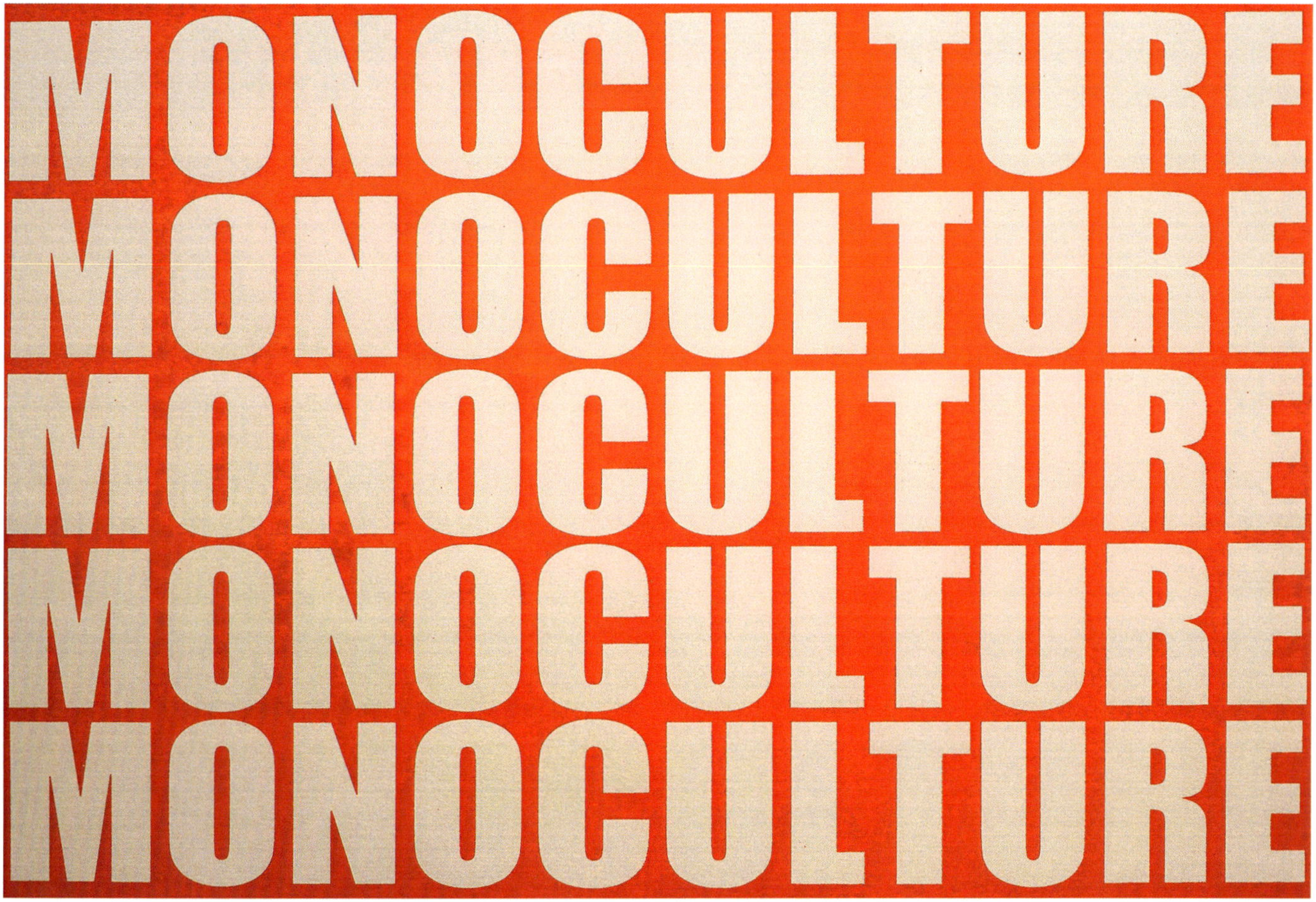

Monoculture

Photomontage
screenprint

Acrylic on
Snowdon

50 x 70cm

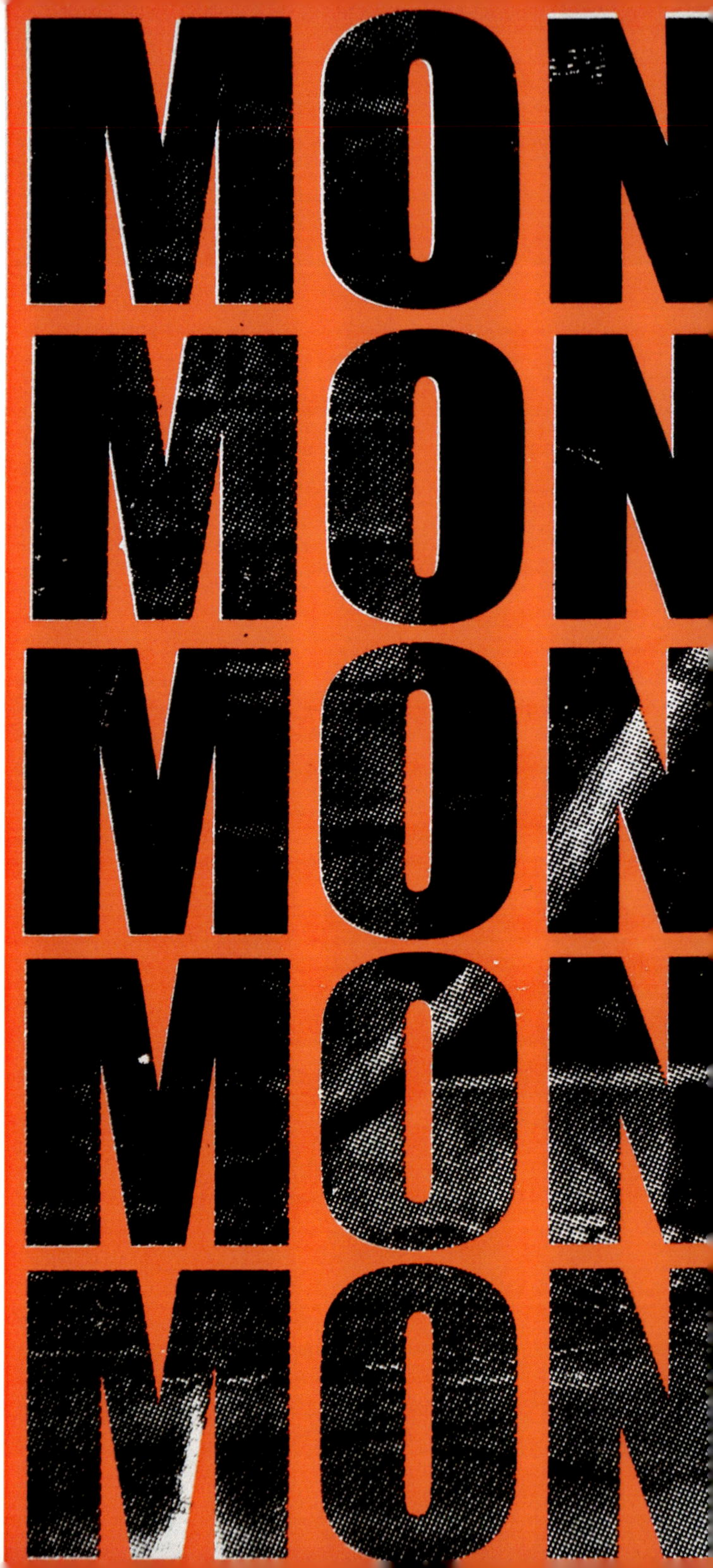

Monoculture

Photomontage
screenprint

Acrylic on Fabriano
Rosaspina
50 x 70cm

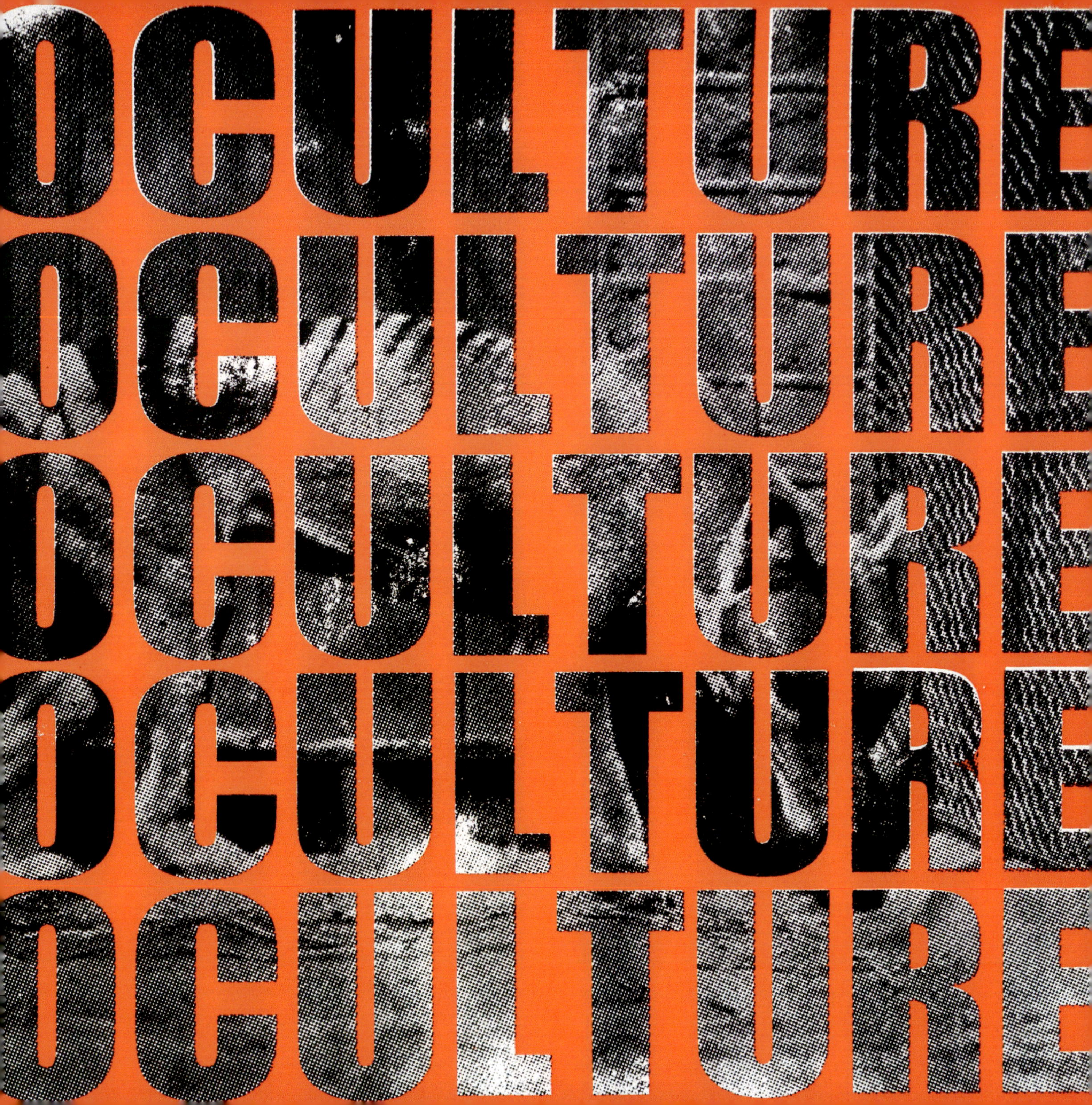
OCULTURE
OCULTURE
OCULTURE
OCULTURE
OCULTURE

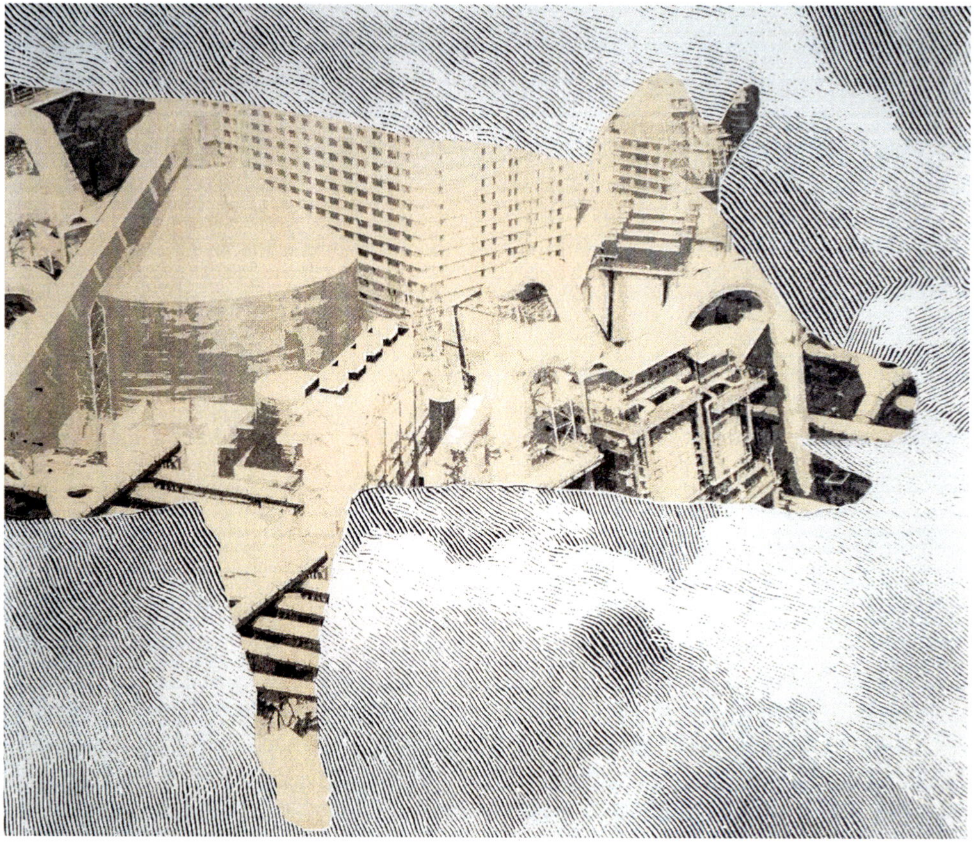

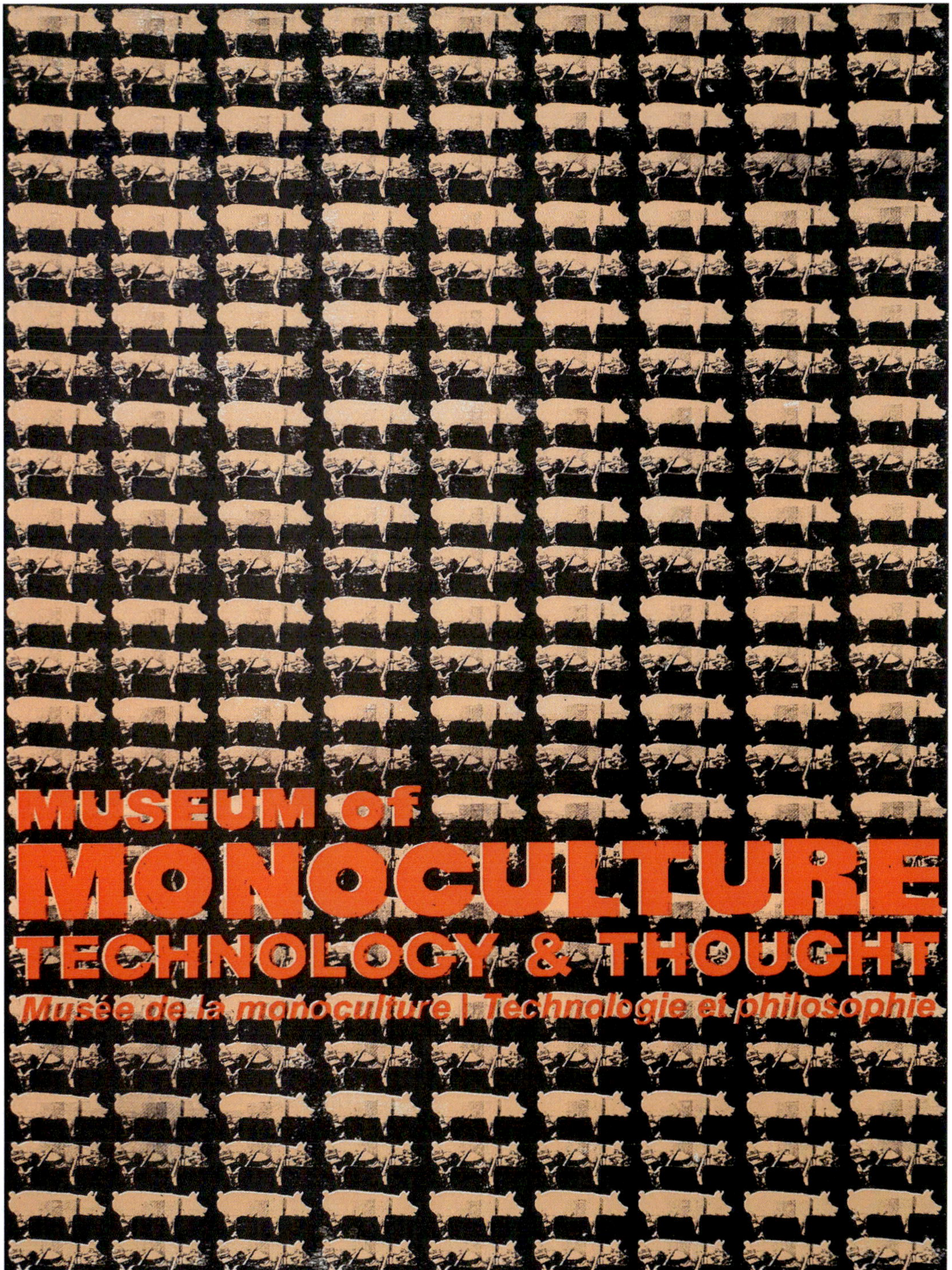

Museum of Monoculture

Photomontage screenprint

Acrylic on Fabriano Rosaspina
50 x 70cm

Pig Swallows Factory Tower

Photomontage screenprint

Acrylic on Fabriano Rosaspina
50 x 70cm

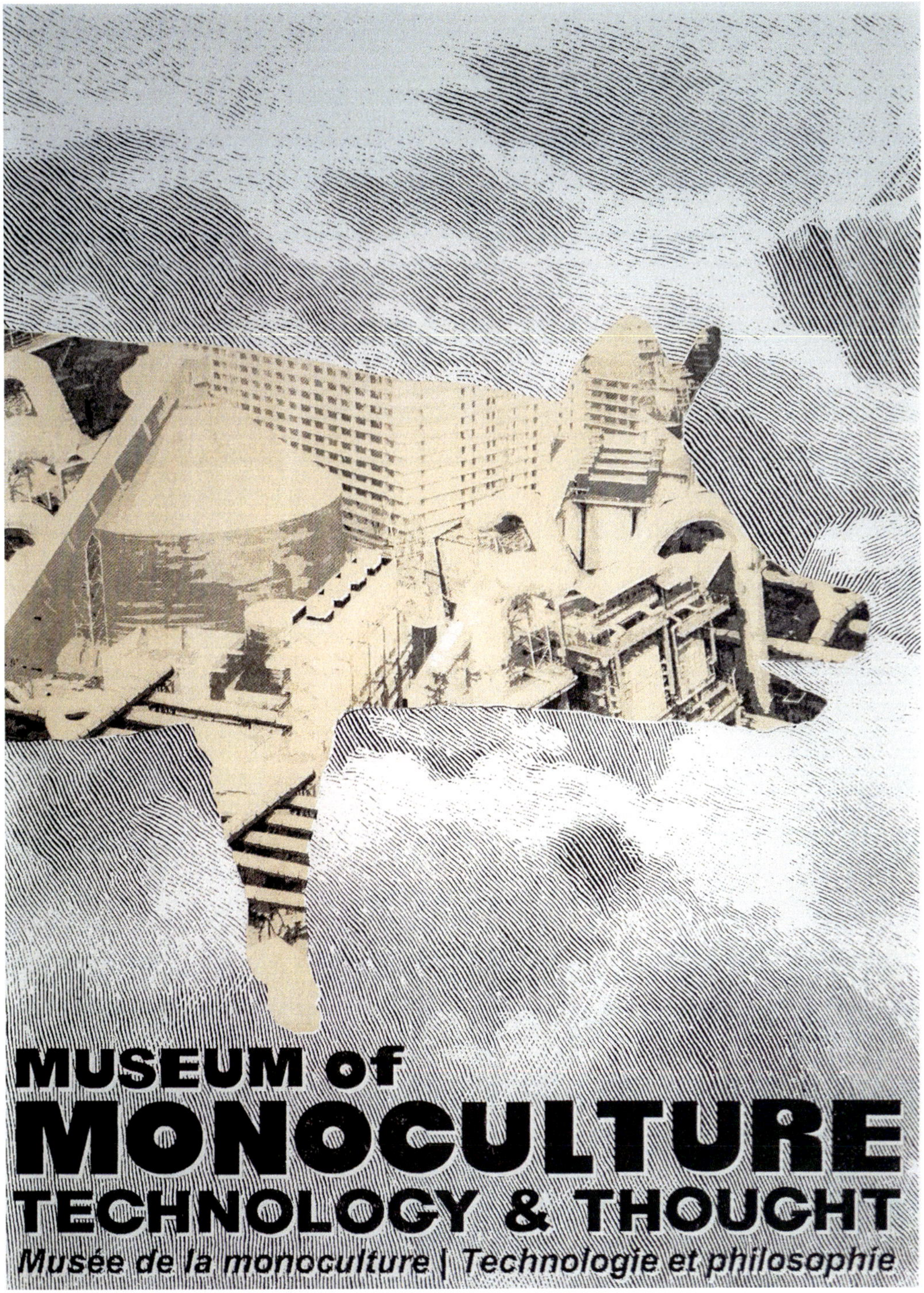
MUSEUM of
MONOCULTURE
TECHNOLOGY & THOUGHT
Musée de la monoculture | Technologie et philosophie

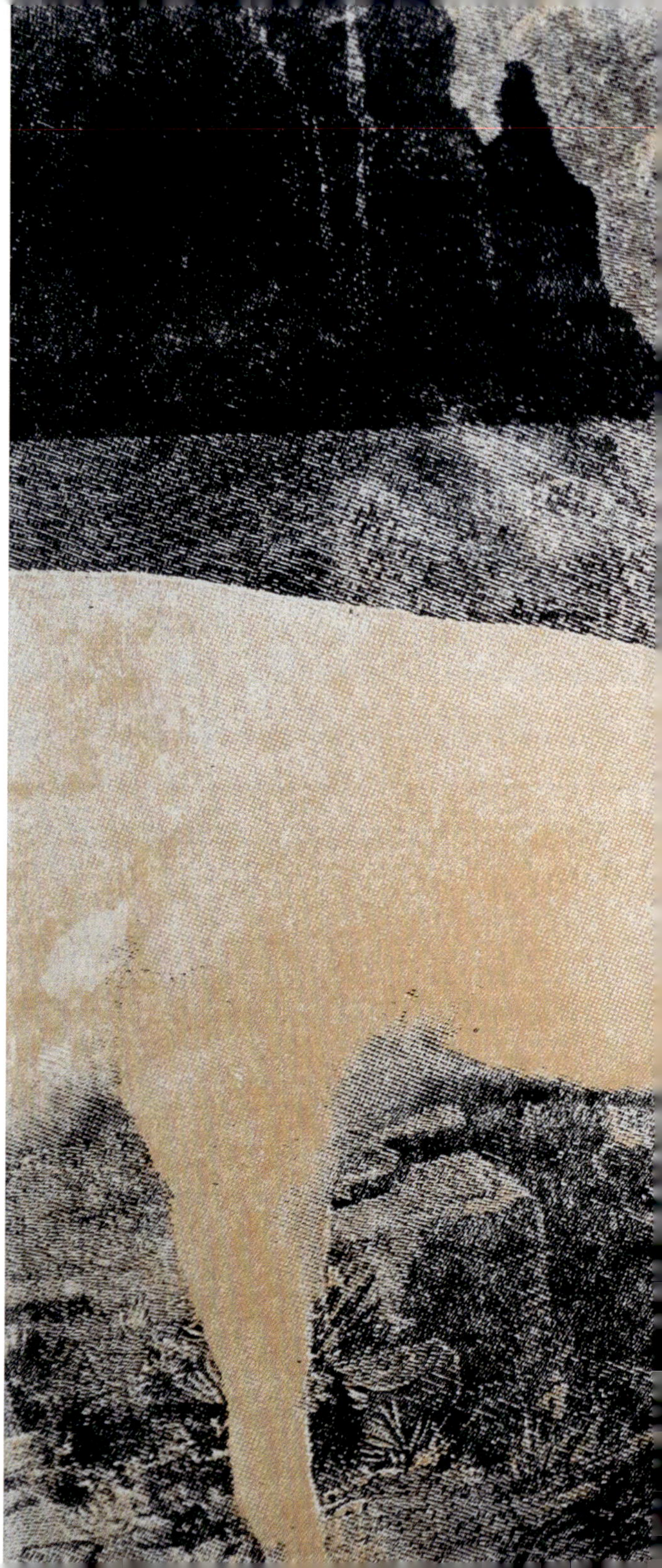

Merci Pig

Photomontage screenprint

Acrylic on Fabriano
Rosaspina
50 x 70cm

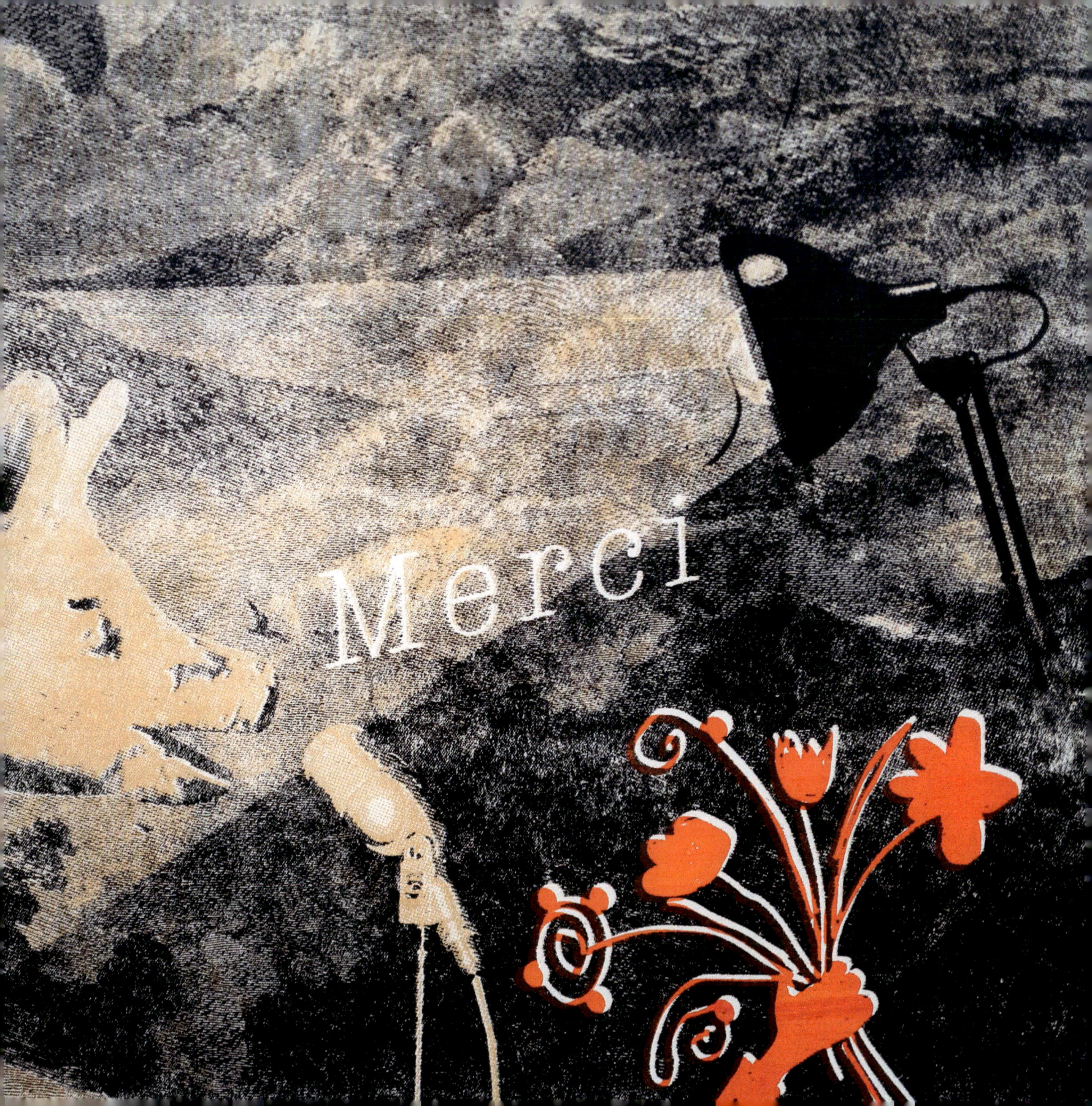
Merci

Australia

Photomontage
screenprint

Acrylic on Fabriano
Rosaspina
50 x 70cm

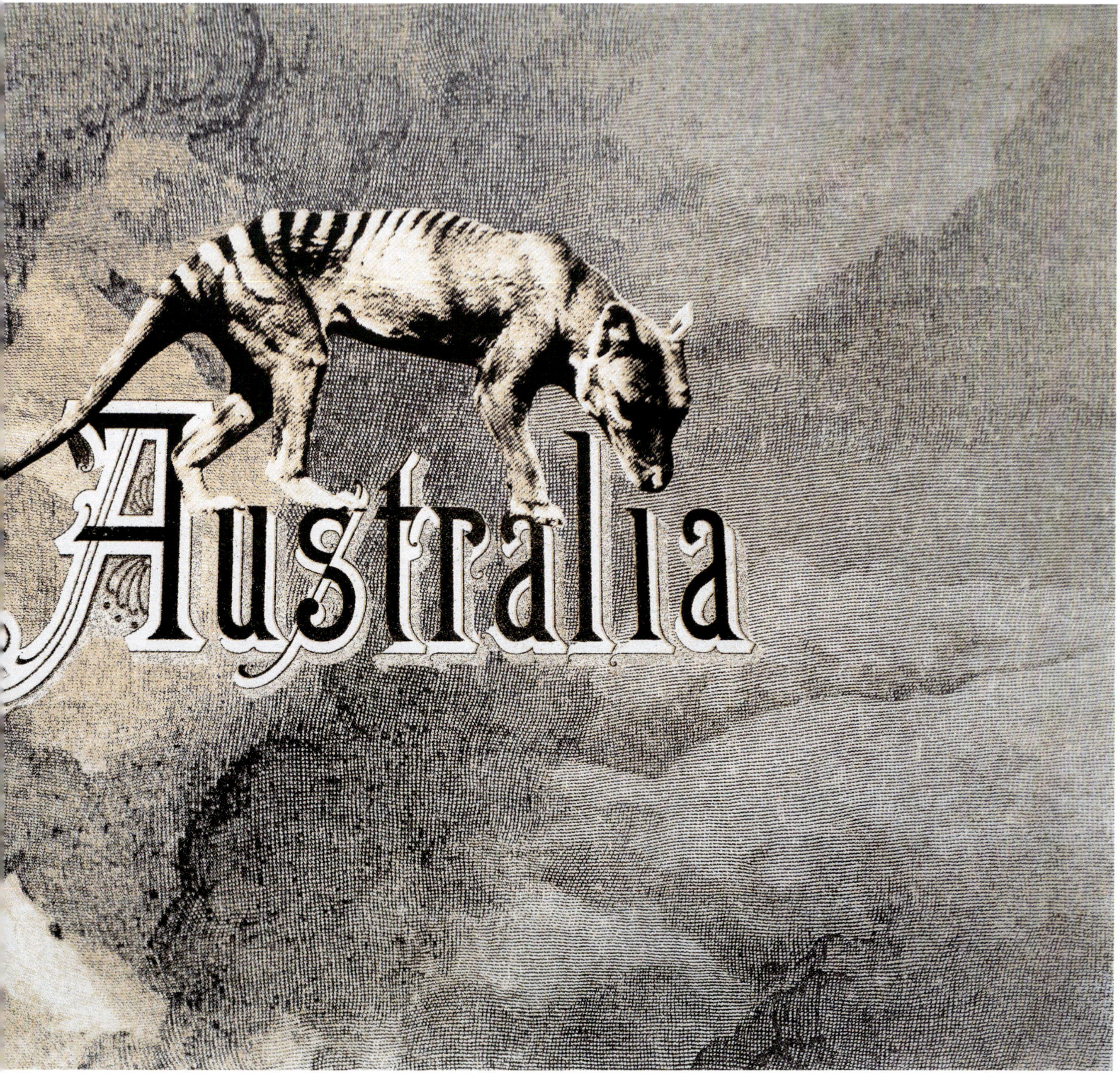
Australia

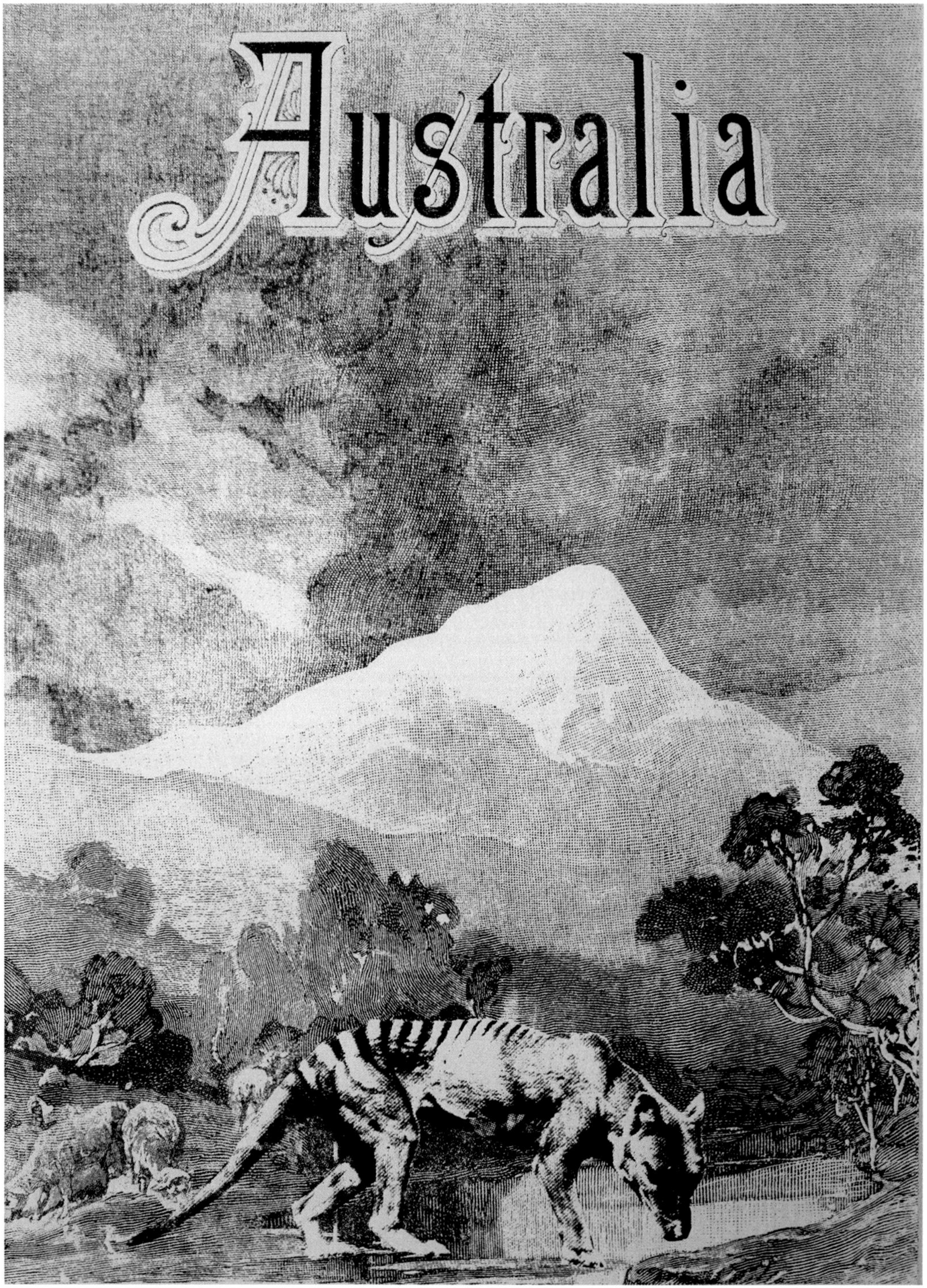

Thylacine

Photomontage
screenprint

Acrylic on Fabriano
Rosaspina
50 x 70cm

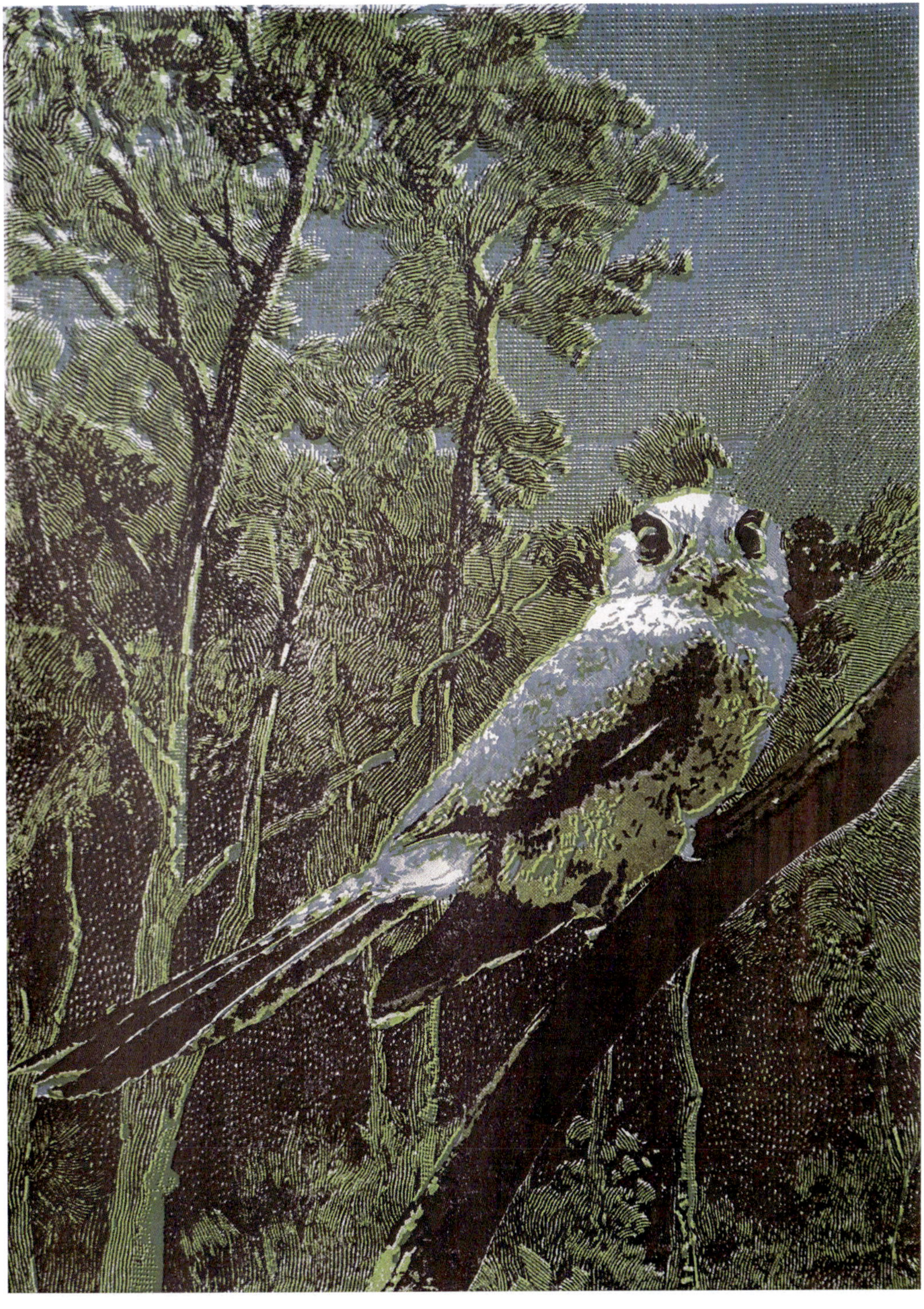

Owlet Nightjar

Photomontage
screenprint

Acrylic on Fabriano
Rosaspina
50 x 70cm

[Brolga] sometimes appeared to collapse in mid-air, and simply fall ...

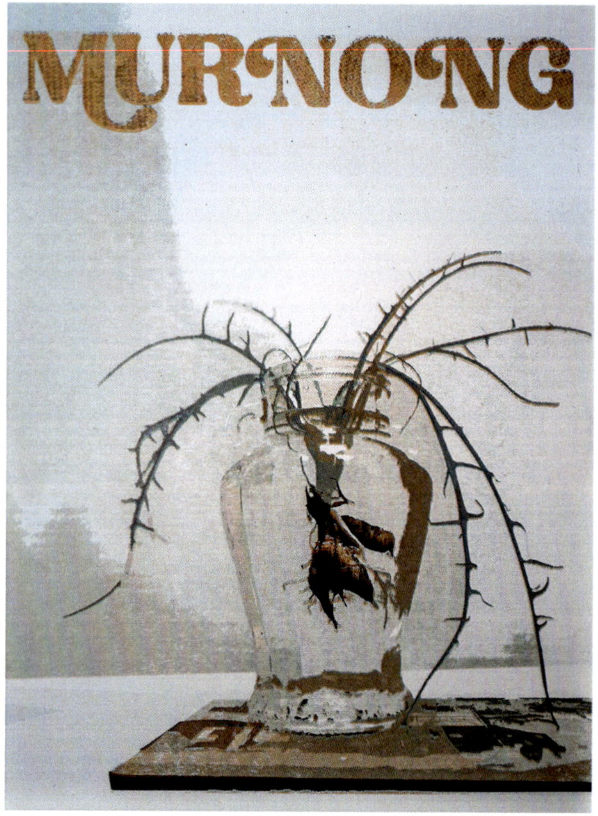

Murnong

Photomontage screenprint

Acrylic on Fabriano Rosaspina
50 x 70cm

THE VANISHING BROLGA

The poem (right) by Richard Bassett, was published in *Colac Reformer* 14 February 1949, he writing:

> I have never seen the (brolga) in great numbers, but an old friend once told me that, in addition to the dance, one of the most interesting things to be seen was a flock of these huge birds in flight. When they decided to alight, they sometimes appeared to collapse in mid-air, and simply fall for a great distance, apparent bundle of feathers and long legs, until they pulled themselves together for the landing, which they made quite unconcerned.

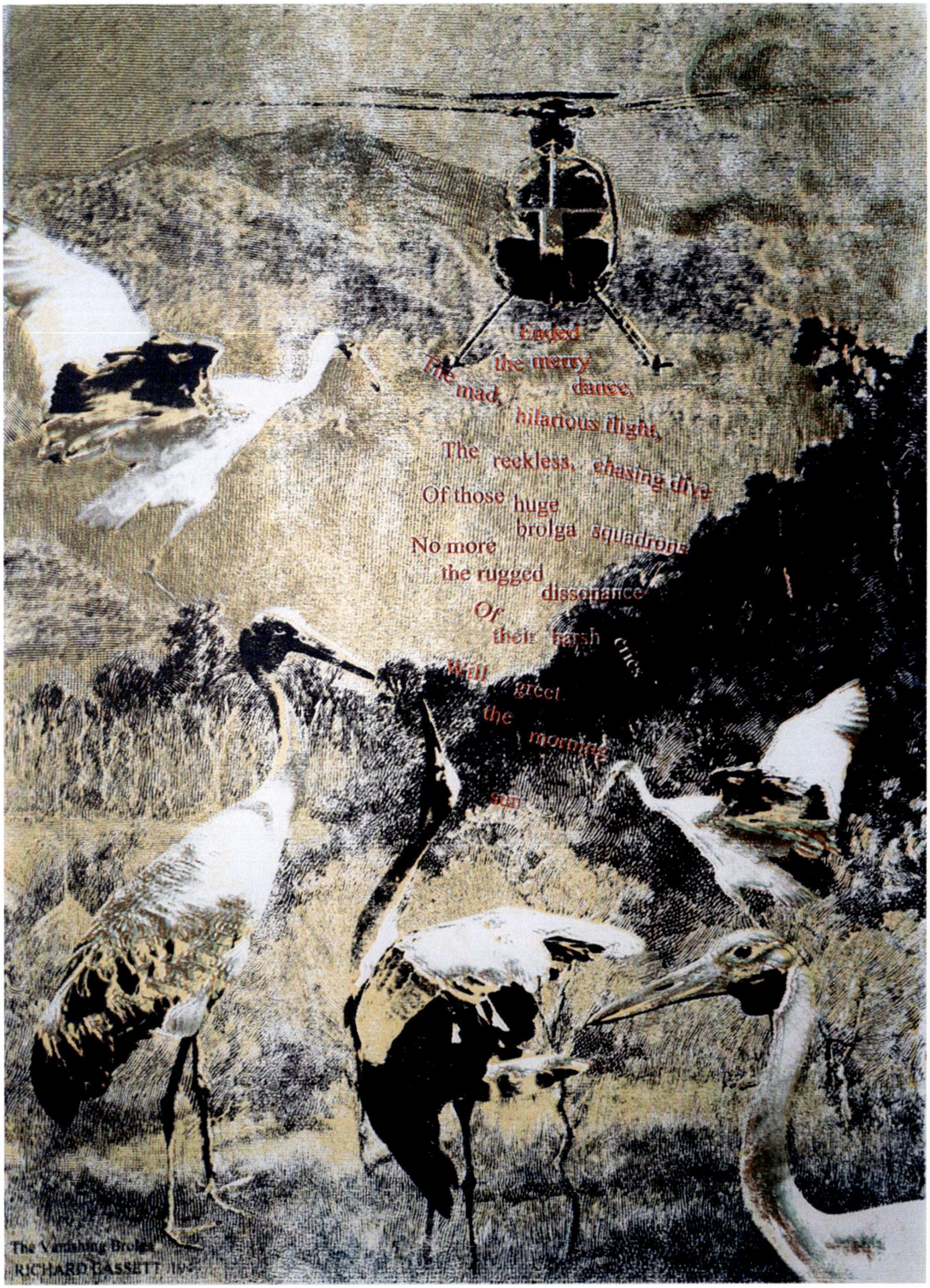

Vanishing Brolga

Photomontage
screenprint

Acrylic on Snowdon
50 x 70cm

... to see a flock of these birds bowing and dancing with that ease and grace

Albert William Morris is the likely author of this article in the Evening News NSW, 1923. Morris recorded his observations of the vanishing wildlife across the Western Division in the 1920s. Brolga, *Grus rubicunda* were once a common sight, travelling in flocks of up to 150 birds across the sky. They were called the native companion, travelling on foot beside the Aboriginal people on their daily food quest. They shared a favourite staple food - murnong, the yam daisy tubers.

In central Victoria, the murnong fields had stretched uninterrupted for miles. Over 100 species of native orchid thrived in amongst the yam daisies and native grasses. Many of these orchids could only feed and reproduce thanks to symbiotic relationship with their specific companion species of fungi – mycorrhizal fungi, that was distributed at scale by the bandicoots and bettongs that lived amongst the murnong fields: it was a highly specialised ecosystem. As the murnong fields were rapidly replaced with sheep pasture and wheat fields, their ecosystem collapsed. Brolga are endangered in Victoria with less than 1000 remaining.

The photograph of the brolga is from the National Archives (1946) with a detail from an 1880s intaglio etching of the Yarra in the background.

THE BROLGA.

Writes A.W.M.:—Out on the plains may be seen the brolga, commonly called the native companion. I have never known fewer than two to be together and have seen as many as 60 in a flock. Between foxes, poisoned baits, and farmers taking their toll, they are gradually becoming extinct. Although insects, etc., are their natural diet, they make short work of the newly-sown wheat fields. They are very timid but will live in captivity. They are fast walkers and exceedingly strong fliers. It is a splendid sight to see a flock of these birds bowing and dancing with that ease and grace that would even put our modern dancers in the shade. Before getting on the wing they bounce a couple of times like a huge windball, and when they set out to fly any distance they rise to great heights, sometimes out of sight, and their presence is only known by their call. While their legs and neck are very much longer, the body is not quite as big as the ordinary farmyard turkey, so some idea can be formed as to the altitude they attain. Their eggs are very prettily marked, are bigger than turkey eggs, and are much sought after, but their nests are hard to find.

Brolga 1923 (2025)

Photomontage
screenprint

Acrylic on Snowdon
50 x 70 cm

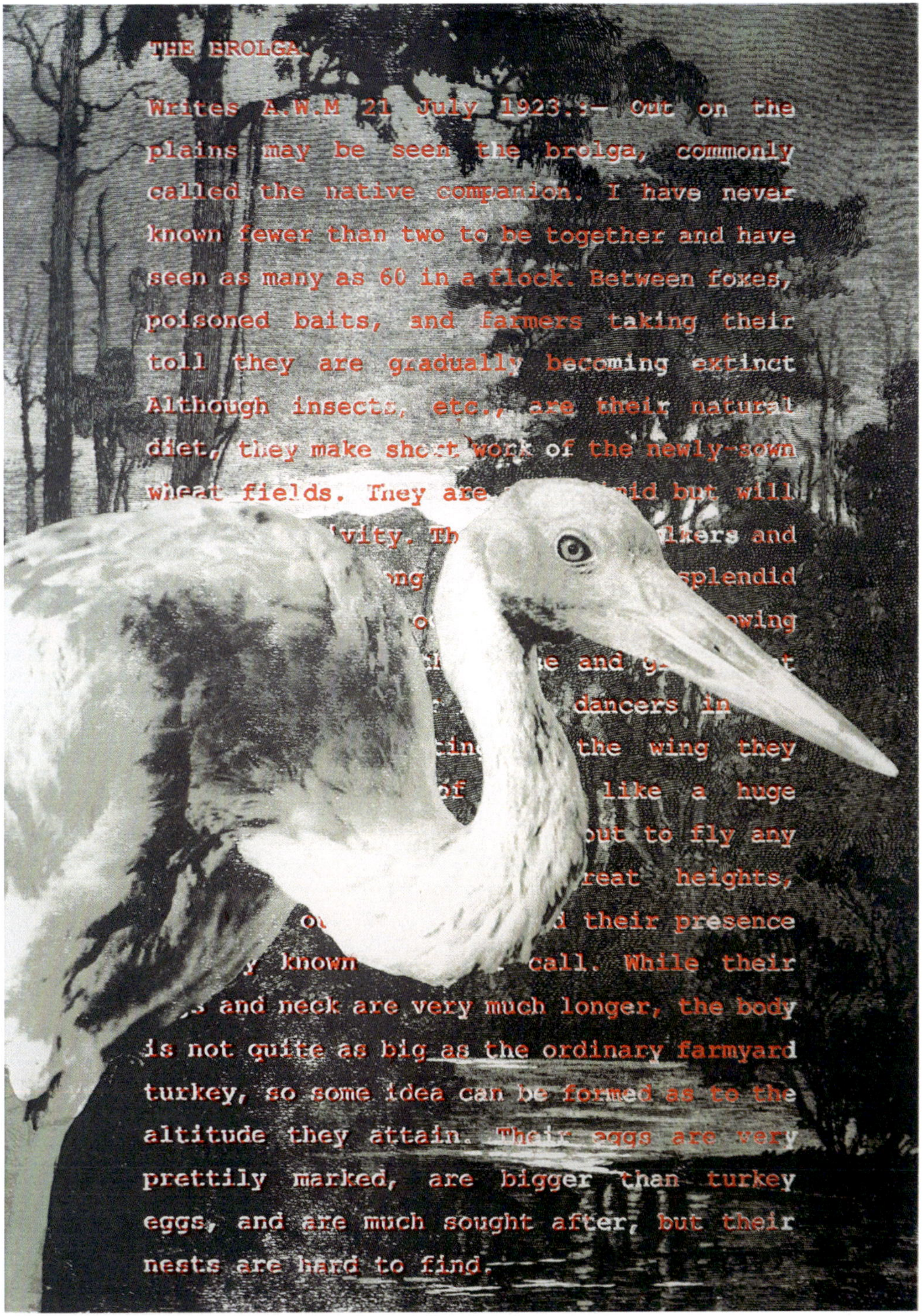
THE BROLGA.
Writes A.W.M 21 July 1923.:- Out on the
plains may be seen the brolga, commonly
called the native companion. I have never
known fewer than two to be together and have
seen as many as 60 in a flock. Between foxes,
poisoned baits, and farmers taking their
toll they are gradually becoming extinct
Although insects, etc., are their natural
diet, they make short work of the newly-sown
wheat fields. They are
kers and
splendid
e and
dancers in
the wing they
like a huge
to fly any
heights,
their presence
known call. While their
and neck are very much longer, the body
is not quite as big as the ordinary farmyard
turkey, so some idea can be formed as to the
altitude they attain. Their eggs are very
prettily marked, are bigger than turkey
eggs, and are much sought after, but their
nests are hard to find.

According to folklore, this poem was composed by a homesick fisherman for his young wife in the first century CE. He inscribed the words on a silk ribbon and concealed it inside a carp, sent as a gift. The simple verse read:

Eat well to keep fit, missing you,
and forget me not.

Carp is one of the most popular fresh water table fish in the world, a staple food in China and Europe. In Australia it is classified a pest species and it is not harvested.

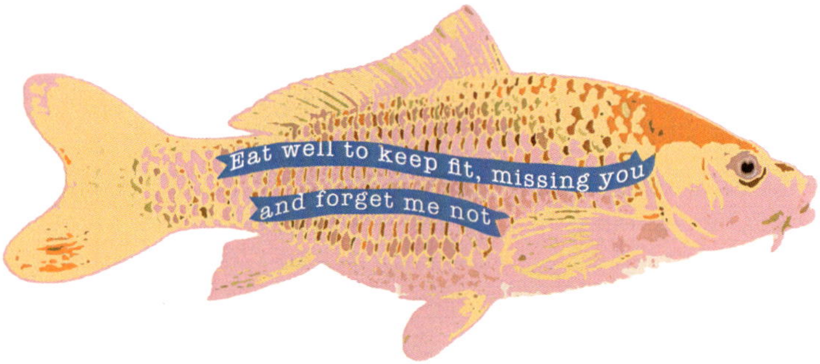

Carp

Photomontage
screenprint

Acrylic on Fabriano
Rosaspina
50 x 70cm

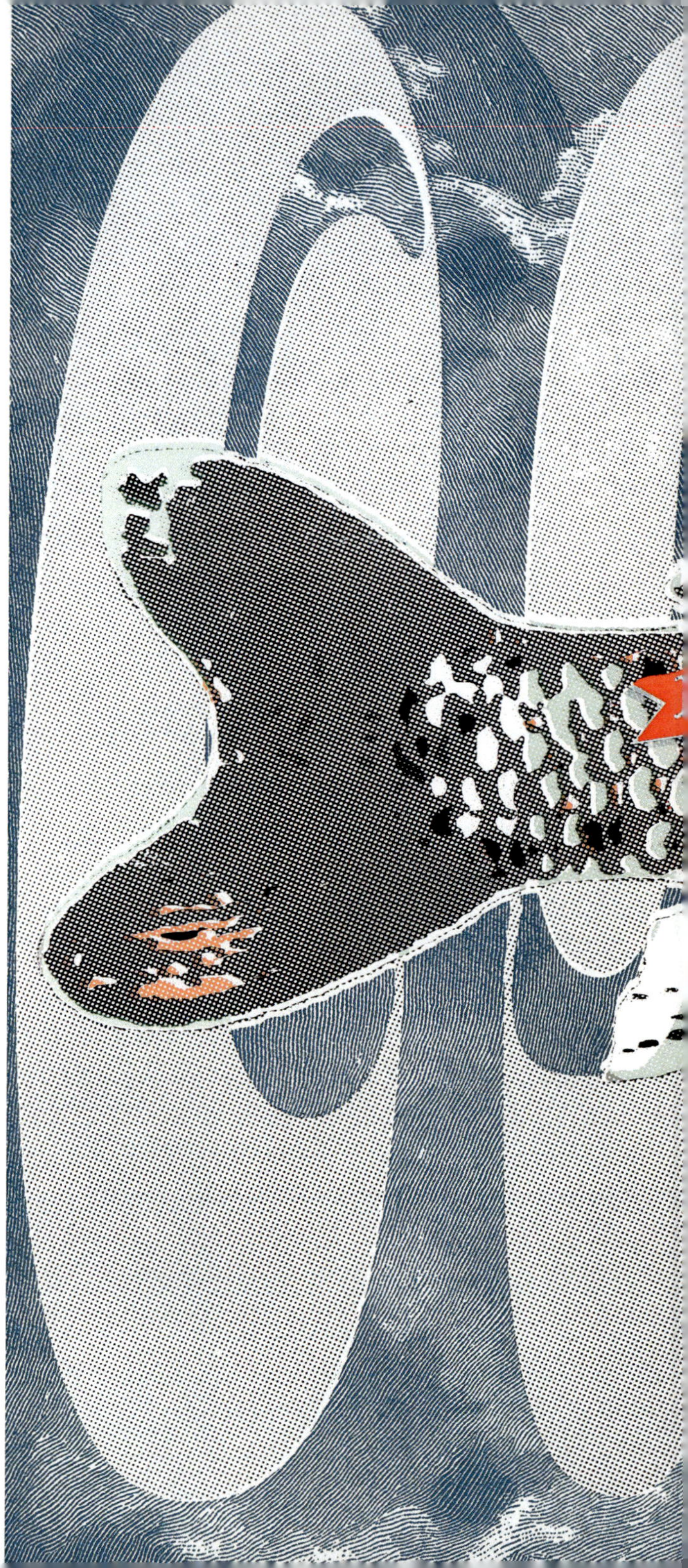

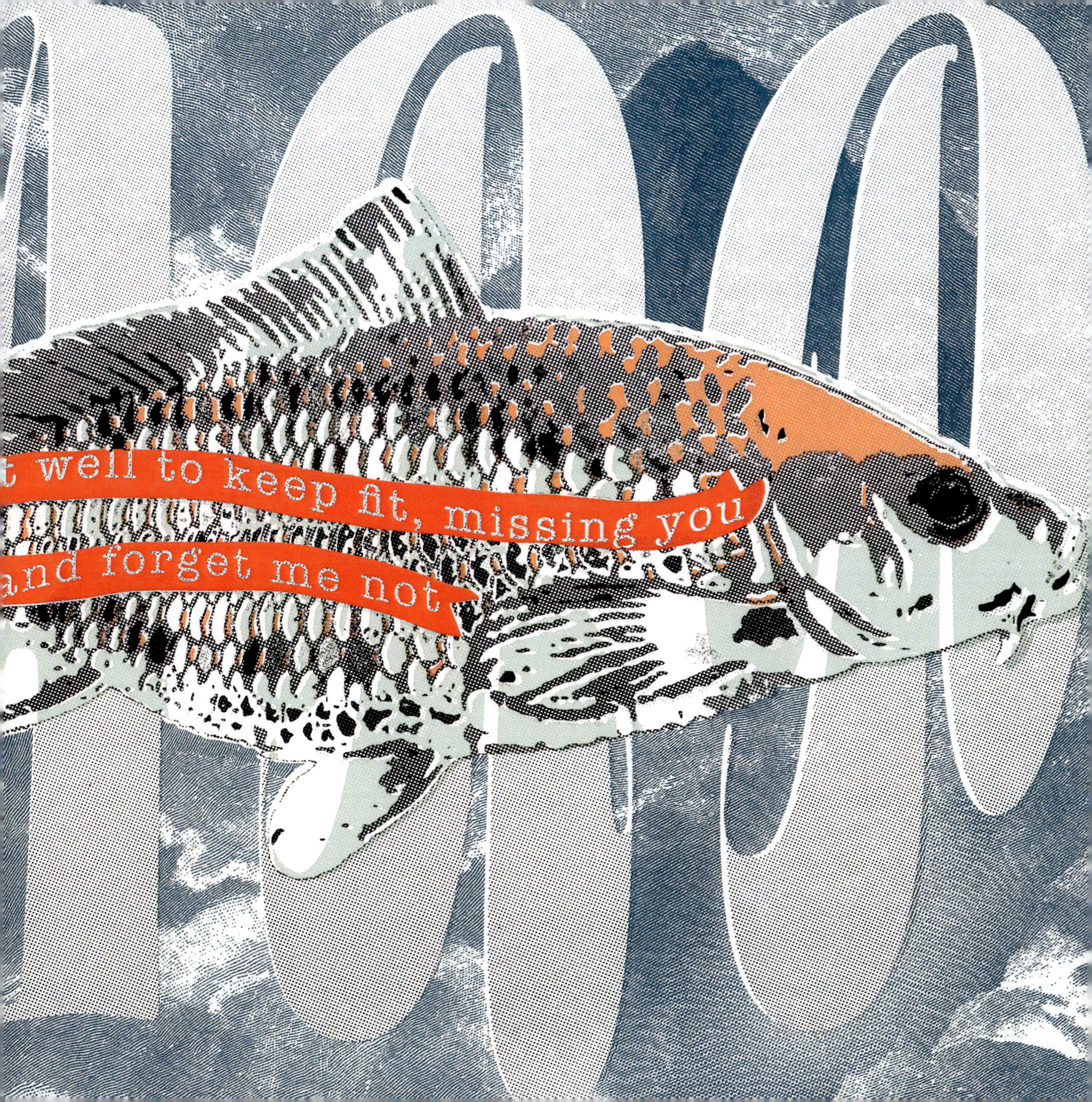
t well to keep fit, missing you
and forget me not

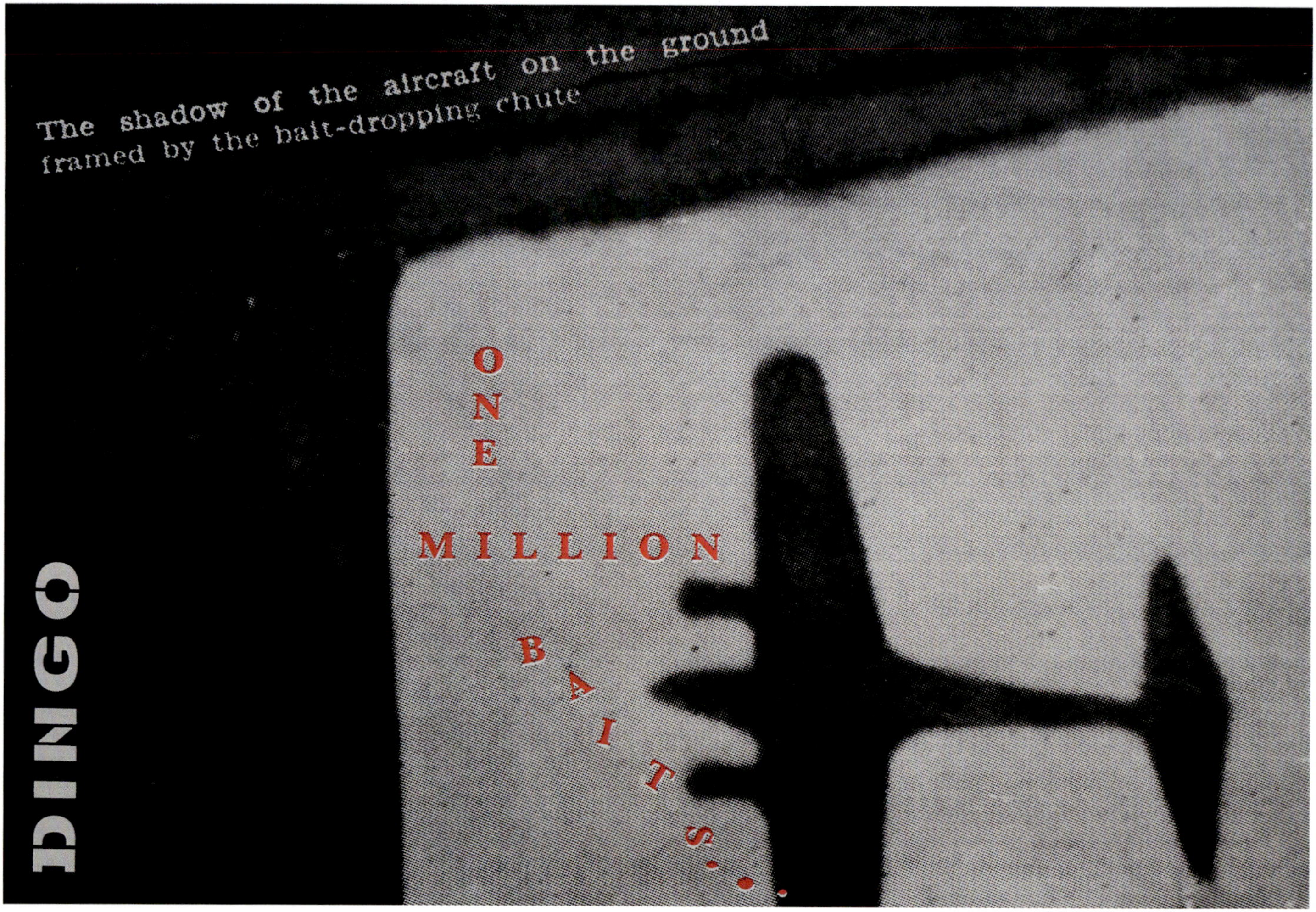

One Million Baits

Photomontage
screenprint

Acrylic on Fabriano
Rosaspina
50 x 70cm

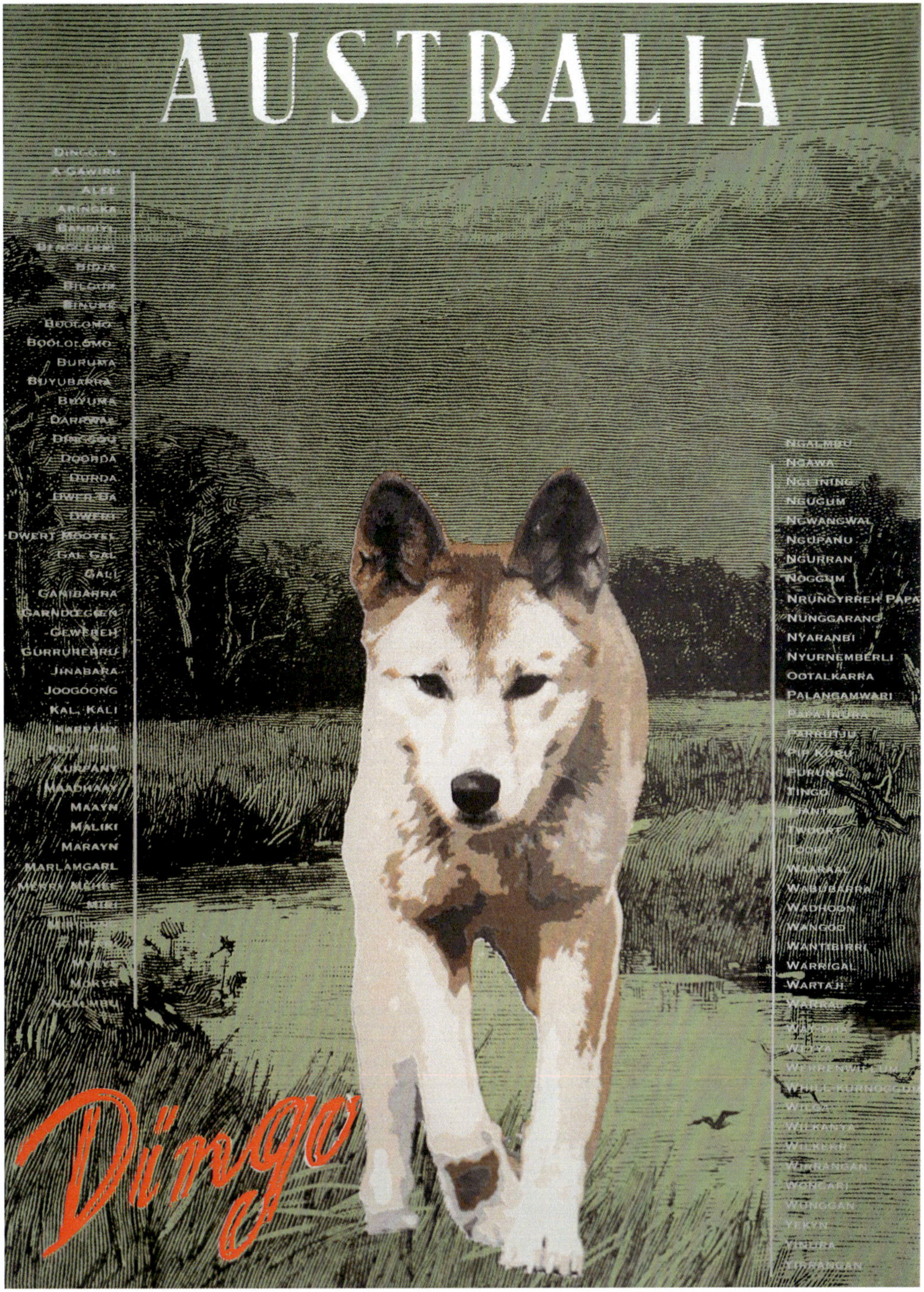

Dingo

Photomontage
screenprint

Acrylic on Fabriano
Rosaspina
50 x 70cm

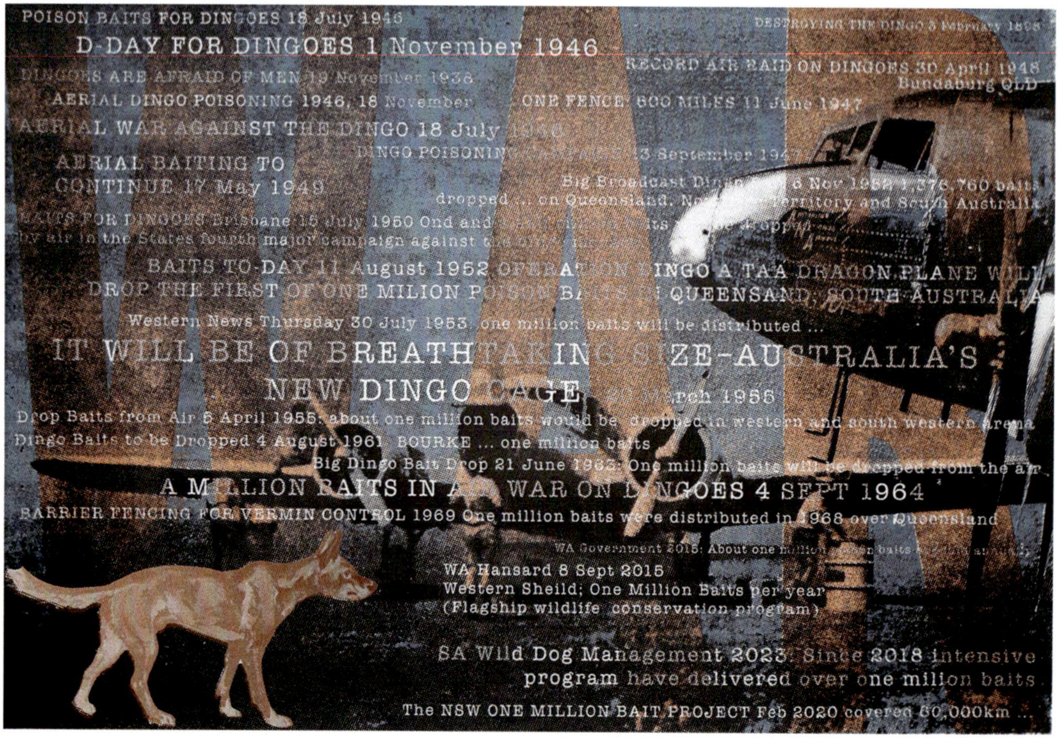

War

Photomontage screenprint

Acrylic on Fabriano Rosaspina
50 x 70cm

LIST OF ARTWORKS

50/200